£1-50

MAURICE BURTON'S

The Daily Telegraph

NATURE BOOK

MAURICE BURTON'S

𝕿𝖍𝖊 𝕯𝖆𝖎𝖑𝖞 𝕿𝖊𝖑𝖊𝖌𝖗𝖆𝖕𝖍

NATURE BOOK

DAVID & CHARLES

NEWTON ABBOT   LONDON

NORTH POMFRET (VT)   VANCOUVER

ISBN 0 7153 7078 2

Library of Congress Catalog Card Number 75–10701

Set in 12 on 13 pt Bembo
Photoset and printed
in Great Britain by
REDWOOD BURN LIMITED
Trowbridge & Esher
for David & Charles (Holdings) Limited
South Devon House   Newton Abbot   Devon

Published in the United States of America
by David & Charles Inc
North Pomfret Vermont 05053 USA

Published in Canada
by Douglas David & Charles Limited
132 Philip Avenue North Vancouver BC

CONTENTS

PREFACE

When I was invited to call on the editor of *The Daily Telegraph* to discuss writing a short weekly Nature Note, the first being accepted and published four days later, I did not anticipate that in 26 years' time I should still be writing these weekly notes. Now I look back on contributions totalling over 300,000 words or the equivalent of four full-length books. Such statistics mask the personal pleasure I have gained from the task, the new knowledge I have accumulated and the number of 'pen friends' I have made as the result of a postbag averaging between six to eight letters a week, as well as phone calls, throughout that time.

Many of the letters contained questions which, as often as not, sent me to reference books for further information or drove me to other sources in quest of explanations. With rare exceptions, all were from people with no professional knowledge on natural history but with enquiring minds and an interested eye on the wildlife of the garden or the immediate neighbourhood. A significant number contained original observations that found their way into my files and caused me to follow modest trails of discovery, so enhancing my own studies and my store of knowledge.

Since the late 1940s and the 1950s saw the phenomenal rise of bird-watching, it was inevitable that the majority of my contributions related to birds. Indeed, I have been told bluntly by a number of readers that when the weekly article was not about birds they did not bother to read it.

It has been flattering to hear from time to time from readers of my Nature Notes who regularly cut them out and added them to their scrapbooks. From time to time also, for some years past, I have been asked why I did not publish them in book form. At last I succumb to the temptation to do this, and in the chapters that follow selected Nature Notes, slightly edited in some instances, are grouped under appropriate headings.

I express my gratitude to *The Daily Telegraph* for permission to do this; to the readers who have encouraged me to do so and to those whose observations and notes are incorporated; and to the publishers, David & Charles, for the opportunity to carry this out.

KITCHEN-SINK NATURALISTS

AT THE BIRD-TABLE

We should be hard put to it to say where the idea of nailing a piece of board on top of a pole to make the first bird-table originated. The credit may belong to someone in Ancient China, where the first organized zoo was established, or this simple pioneering event may have occurred in Germany, Austria or Switzerland. The United States has the reputation of having entered early and enthusiastically into the hobby of caring for wild birds. It may be taken for granted, perhaps, that the bird-table was evolved where domestic cats were a menace, indulging their horrible habit of lurking in shrubberies and borders to pounce on birds feeding on the ground. It is sufficient that whoever was the unknown pioneer, the result of his or her brainwave was to give pleasure to millions.

In the early years of this century, even before, there were ornithologists proper. They were a select band, mainly amateurs, dedicated to the study of birds and often looked upon with something approaching contempt. After World War I their practices spread to give birth to the pastime of bird-watching, one of the outstanding social phenomena of this century. Their numbers were later augmented by even more numerous and less erudite but no less enthusiastic devotees. One of them described herself to me as a kitchen-sink bird-watcher. She and her like form an army that could not have come into being without the use of the bird-table.

As the kitchen-sink watchers multiplied so did the devices for encouraging birds. The tables themselves became more elaborate

Cats and blue tit

or were festooned with half coconuts, wire spirals, peanut bags and the rest. The birds responded and more and more species joined the parties.

So we give the bird-table pride of place in this book, and especially because, to many who read it, interest in the subjects of the other chapters will have started with or been stimulated by what had been seen at the bird-table.

YEARS OF FAT

Small birds like blue tits, great tits and coal tits reap a rich bounty from their natural gymnastic abilities in dealing with food suspended from the bird-table. The mere fact that they are entertaining leads us to keep them well supplied. But one discerning reader, in the Sevenoaks district in Kent, has been presented by them with a problem. He makes a habit of putting out coconuts, and he has noticed that in every other year the birds get through about 10 nuts but in the alternate years only about two. He remarks that he cannot explain this as conditions each winter are approximately the same.

We know that the numbers of any species tend to vary from year to year. These are normal fluctuations but they are hardly so great as to give a five-to-one difference in the amount of coconut consumed. Nor would they give this difference with the regularity implied by my correspondent's statement.

It may be there is an interesting key to the puzzle. These birds

Blue tit at window

feed largely on insects during the summer, and they continue searching for them in winter, but then the supply is considerably diminished. So, in the colder months, they turn to berries and other wild fruits, to supplement their feeding, and especially to beech mast.

There are good reasons for believing that the supply of beech mast has a two-year periodicity, abundant one year and very scarce the next.

DEFEATING THE SPARROWS

Those who keep a birds' feeding-tray prefer to see the food being taken by birds other than house sparrows. Sparrows have on other counts as much right to our consideration as birds with more showy colours or something nearer a song; but they tend to hog whatever is going. Often the cry goes up: 'How can I keep the sparrows off my bird-table?'

A solution suggested by the late David Seth-Smith, formerly one of our leading ornithologists, was to stretch a piece of black cotton across one side of the tray, about 3in up from the rim. Most birds, he claimed, do not seem to be bothered by this, but sparrows keep away. Whether they are more suspicious, or whether the trick discovers some weak spot in them, is beside the point.

The beauty of this plan goes beyond its simplicity. It points a possible lesson in controls. There are a number of birds and other animals which give interest or colour to life but, having one or more habits that clash with human interests, are destroyed. How much better if for each we could discover a simple method of finding some Achilles' heel which would keep them within bounds instead of killing them.

Since I have never been much bothered by sparrows I have not tested the method.

THE TIP-AND-RUN MARSH TITS

A visitor once to our bird-tray was a foreigner beside the usual crowd of blue tits and great tits. Slightly smaller than the blue tit, its outstanding feature was a neat glossy black cap, contrasting strongly with the rest of the plumage. The underside of the body was buff, the back and wings brown and the cheeks white. There was also a small black bib on the chin. It was the marsh tit.

Another species, the willow tit, found almost exclusively in south-east England, is very like the marsh tit. There are slight differences in colour and in the call, but the best identification mark is, perhaps, the white patch on the wing of the willow tit.

Whereas a dozen other kinds of birds were scattered over the garden right up to the house and visited the feeding-tray outside the window in turn, the marsh tit showed a very set behaviour. From a perch in the tree on the edge of the garden it flew direct to the feeding-tray, grabbed a piece of food and flew back to the tree.

The frequency of its visits and the amount of food taken away suggested an incredibly large appetite until we saw that there were at least two of them, following exactly the same course. They were strangers, in our world, making tip-and-run raids on the feeding-tray.

THE LONG HAUL

A net bag filled with peanuts and hung in a new place in the garden, on a stag's horn shrub, quickly attracted a collection of birds. The tits found it first, and soon sparrows were gathering on the ground beneath to pick up fragments dropped by the tits. Then greenfinches joined in, and later a solitary goldfinch was seen clinging for a short while, tit-like, to peck at the nuts.

Goldfinches can perform another trick for which tits are apt to be given exclusive credit: pulling on a string to obtain food. Whether they ever do this with peanuts threaded on a string is unlikely, but their ability to do something like it has been known for a long time.

In the sixteenth century a goldfinch was called draw-water, or its equivalent in several European languages. It was then one of the favourite cage birds, and it was the custom to keep it in a cage so designed that to survive it had to learn to pull up a tiny cart containing seed on the end of a string on one side of the cage. The idea

Goldfinch

13

was that its owner could sit and watch. The bird would pull the string with its beak, hold the loop with one foot, then pull in another loop, hold that, and so on, until it could take the seed.

On the other side of the cage was a thimble full of water, also held by a string. To drink, the bird had to draw this up in the same way.

LEARNING FROM HUMANS

In 1968 I wrote of greenfinches developing a new habit of coming to the bird-table and taking peanuts. I had been impressed by the number of people who, in a space of a month or two, had told me they had noticed this.

However, I once wrote about nuthatches having started to come regularly to bird-tables in the early 1960s. Then a reader wrote to say that she remembered seeing nuthatches at bird-tables back in the 1920s. Perhaps we should be cautious also about calling this a new habit in greenfinches. It would probably be more correct to say that what the greenfinches are doing is responding to a new

Greenfinch

14

habit in human beings. Considerable ingenuity has been exercised in evolving and marketing new gadgets for feeding birds, such as the net bag and the wire spiral in which shelled peanuts can be hung.

Greenfinches have wider feeding habits than most finches. They can feed readily in trees and bushes or on the ground. They take yew berries, blackberries and hips and, lower down, take the seeds of many weeds, as well as cereal grains. Regrettably they sometimes take fruit buds, but they also take insects, including aphides. Taking peanuts may therefore have been only a matter of extending their feeding habits to include a new gadget.

BIRDS OF A FEATHER . . .

From 100 letters received in early 1968 about my notes on greenfinches it is clear that the habit of taking peanuts started in some places several years before and had since been building up, but that in other places it had only recently started. It also emerges that other birds, including sparrows, robins and chaffinches, tend to follow suit. The natural question is whether these others are copying the greenfinches.

As with many other animals, most birds seem to be attracted to their own kind, especially when these are feeding. This is referred to as social feeding. It is particularly marked in the domestic fowl: the chicks begin in their first 24 hours to peck at anything the hen pecks at, so learning what food to take.

The chick instinctively pecks at the hen's beak or even at a beak on a model of a hen's head. When a pellet is attached to the end of the model's beak the chick pecks at that. Consequently, a chick standing near a hen as she lowers her head pecks automatically at anything she pecks. When a mechanical model of a hen was made to peck at grains of two different colours, chicks near it usually pecked at the same colours as the parent-model selected.

Tests on chaffinches and sparrows indicate that they seem to be attracted to join even other species, when they are feeding. They also show that the young birds of both these species will feed not only on familiar foods but on unfamiliar foods as well, so long as they have seen other birds doing so first.

INSTRUCTIVE SCRAPS

A noted and showy visitor to the bird-table is the great spotted woodpecker. People ask what food to put out to encourage it further.

Half a coconut or fat will prove attractive to it but, above all, grated cheese is the great draw. There need be no waste about this; rinds or stale cheese are perfectly acceptable. Another method we use is to make a basket of small-meshed (half-inch) wire-netting and fix it to the trunk of a tree or a fence. Any odd pieces of fat thrown into this will be pecked through the mesh by tits, occasional woodpeckers and a variety of other birds.

By placing various foods in different positions or situations it is possible, as many of us know, to see much that is entertaining and instructive. It is entertaining to see efforts made by birds that normally feed on the ground to emulate the gymnastic efforts of others, such as tits, as they try to reach foods hung up with no perch handy. It is instructive to see how the normal pattern of feeding still holds when unnatural food is being taken, best illustrated by the nuthatch.

This elegant bird, plump, short-tailed, with a woodpecker-like bill, is blue-grey above, and buff below, with chestnut flanks. It is accustomed to hammering at nuts, acorns and hard seeds which it wedges in crevices in bark to open the shells. When feeding on soft cheese it deals the same sledge-hammer blows.

AIRBORNE COURAGE

Birds, and other animals for that matter, have all kinds of built-in mechanisms for keeping members of their species spaced out, and for keeping the different species from coming into competition with each other. The free-for-all so often seen at the bird-table gives a fair insight into the turmoil there would be if things were otherwise.

In these conflicts victory is not always linked with size, or the starlings would not be allowed to boss as they do. They profit from the characteristics of standing their ground and laying about them with a dagger-like beak. Even the blackbird, another bully, gives way to them, and it has to learn to do so. Young blackbirds will go for a starling, but only once. The lesson is usually permanent.

Blackbirds take it out of thrushes, however, and not only at the bird-table. It is quite remarkable how a blackbird seems to know the precise moment when a thrush has beaten a snail's shell to bits and is ready to swallow the succulent body. It will then rush in and grab it.

Let the two birds take to the air, however, and the roles are reversed. The thrush then chases the blackbird. Indeed we need not go to anything so large for this reversal, because off the ground even a chaffinch, seemingly so docile, will make a dead set at a blackbird. And an airborne blackbird will even chase a starling.

PERENNIAL PROBLEM

Should we remove the nesting material from the nesting boxes used by tits? This is a little aside from our subject but bird-tables and nesting-boxes go almost hand-in-hand. In the early months of the year I am often asked this question. Some of those who ask it have been proffered advice which conflicts with their own inclinations. Others have received conflicting advice. It is one of those questions which any one person can only answer from his own experience—and then stand by to be contradicted.

My own inclination is to leave the nesting material alone. I have found, over the years, that both blue and great tits may do one or the other of two things. Some merely build on top of the old nest, and others meticulously remove all the old material, taking several days to do so. They scatter it on the ground all around; then they start to rebuild and spend the next few days picking up every scrap they have carried out, returning it to the box.

From what I can remember blue tits are more inclined to build on the old nest than great tits. That this can have few deleterious effects is shown by the way one can sometimes find anything up to half-a-dozen nests in a pile.

Great tits at a coconut

The more one looks into the matter of nest-building the more one is inclined to the view that birds, like us, have diverse tastes in the matter of furnishing and of spring-cleaning. Some people might argue that removing the old materials also removes the parasites left in from last year. There may be some truth in this, but on the whole I prefer to do nothing, on the assumption that the birds know best. After all, tits have nested in cavities in trees, for which the nesting-box is an artificial substitute, for longer than man has been civilized.

TIGHTROPE PERFORMER

There is one interloper that interferes more than any other with the smooth working of the bird-table. This is the grey squirrel that should serve, were it not so troublesome, as a symbol of Anglo-American unity, because it was imported from the United States. It was brought in as a pet on several occasions towards the end of the nineteenth century and misguidedly liberated. At first nothing untoward happened and then the population explosion took place. Today it ranges destructively over most of Britain.

Many people have tried to defeat and outwit the grey squirrel, that trespasses on their bird-table. A common practice is to stretch a rope or a stout wire between two posts and hang the food intended for birds in the middle of it. Even this fails. On several occasions during past years people have written to me about squirrels walking the tightrope. Usually a wire has been stretched between two points at about head height, with a half-coconut suspended by a string in the centre of the wire, a dodge intended specifically to defeat squirrels. Sometimes I have been told the squirrel walked along above the wire, at other times that it started to do so but swung below it and continued hand-over-hand.

A few years ago I put up several experimental wires in front of my study window, each with its half-coconut in the middle, in the approved fashion, between trees where a squirrel had a regular itinerary. I never saw the squirrel at its tricks but the coconuts were cut down, the string bitten through.

Later, I received a strip of film from Canada showing a grey squirrel walking on a wire. More interesting is to see it balancing to eat the food. It turns so that its hind feet are on the wire with its hindquarters to one side of it, tail curved underneath, and its head lowered on the other side. In that position its main weight is below the centre of gravity, the essential trick for tightrope walking. There are moments, however, when the squirrel raises its head and sits upright for a while, a piece of superb balancing of which the mechanics are difficult to work out.

BED OR BOARD?

Fieldmice are habitual hoarders and they often use strange places for their caches of nuts, berries and seeds. One sometimes comes across these in the ground, and the compost heap is another favourite place. Old birds' nests are also used, but it is uncertain whether these are feeding platforms or store places. Possibly both.

An old nest of a blackbird or bullfinch may be filled with a consolidated mush of hips, haws and other berries. All the mouse is interested in is the pips. It gnaws these open and extracts the kernels, leaving the pulp and the shells filling the nest to the brim.

A typical example is recorded by a reader, except that this one was in a nesting box. She describes finding the box filled to overflowing with red holly berries. Under the top layer was a layer of dried berries, wheat kernels and other unidentifiable matter. Beneath this the next layer was as hard as a brick and at the bottom of it all was ordinary disintegrated nesting material.

It has been suggested that when a mouse uses an old bird's nest in a hawthorn tree, for example, it uses it as a feeding table, collecting the haws from the outermost twigs and consuming them in the nest. There was, however, a similar accumulation in a nest 5oft up in an elm. This must have meant many long climbs to take the berries up to it. A fieldmouse can jump down from heights of 15–20ft without harm.

SHEDDING SPINES

The hedgehog is about the most familiar of our native animals, and one about which we still need a lot of information. I had been keeping watch for two years in the hope of answering a reader's question.

Then another reader raised the same question in a different form: he told of a hedgehog in the garden, for which he put down milk. On one occasion he noticed, after the animal had gone, a dozen or so spines lying on the ground where it had stood. The question in both cases is whether hedgehogs moult their spines.

I find that they moult them a few at a time, beginning in late April or early May and continuing into early summer at least, sometimes even later. The hedgehog's spines are modified hairs, so we should expect them to be shed just as other animals shed their coats, but they no longer conform to our usual idea of hair. Each is spindle-shaped, tapering to a point at the outer end. At the inner end is a flexible neck, slightly curved, which finishes in a ball joint set in the skin. The surface of the spine is grooved longitudinally and coloured with irregular bands of brown and white.

The flexible neck is a shock-absorber, so that the base of the spine shall not be driven into the skin. A hedgehog may climb a tree or a fence, and it can fall from a height to land on its spines without suffering injury.

CHAPTER 2

SINGING DAY AND NIGHT

LITTLE SONGSTER

A rich, full-throated song burst on the still morning air, when the heat of the sun had driven most birds to seek the shelter of the hedges, there to offer up only occasional feeble notes, or to find refreshment at the edge of some shaded pond. A wonderful burst of melody, exquisite and haunting. A compound of sweet piping notes, of crescendoes and diminuendoes, trills and cadenzas.

A pause, and the same rich melody is repeated a few paces farther on along the hedge. No quiver of foliage has betrayed the movements of the singer. Again the song is trilled forth, a pace farther on, and as we approach, searching the thicket of the hedge for the invisible musician, it bursts forth again, always a pace ahead of us. A quick, cautious approach and we are abreast of the next burst of song; but scanning the foliage at a little distance from it, we catch no sight of the wandering minstrel as he continues carolling down the hedgerow.

Determined to see the melodious will-o'-the-wisp, we push on ahead of him to where the stems and branches of the hedge are but sparsely covered with foliage. Here we lie in wait. Bursts of song draw nearer and nearer, but never a leaf stirs or a twig creaks. Suddenly, almost invisible among the brown tangled branches, moving with a mouse-like silence from twig to twig, a small bird enters the bare spot in the hedgerow. A bird of tiny proportions, with stumpy tail erect, hops on to a twig, pecks at a leaf, looks round with a beady eye, and opens its mouth to pour forth a surprising gush of sweet melody.

22

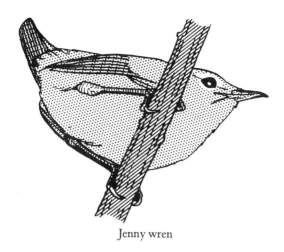

Jenny wren

Who would have thought the tiny wren capable of producing such a volume of sound?

THE FANTASTIC DAWN CHORUS

The surprising thing is that so few people have heard the dawn chorus or know anything about it. Even more surprising is the disbelief expressed when those who have heard it for the first time describe it to their friends.

In a way this is not to be wondered at, for, heard at its best, in late April or early May, its volume is almost fantastic.

As night fades into that mysterious grey preceding the dawn, the first notes are heard. At first scattered, perhaps a pipe or two from a blackbird or robin, the chorus gathers strength until every bird is joining in and the symphony rises and falls. For an hour or more it goes on, then almost as abruptly as a musical piece, it ceases.

The reason for it is not clear. It has been suggested that the energy accumulated during sleep is released during the time of waking and the time when feeding begins. In birds the release of pent-up energy would naturally be expressed in song. The dawn chorus is heard in simple form just before the onset of the breeding season, rises to a peak as that season advances, and begins to flag as the nestlings appear. There is a similar chorus at sunset, though of lesser proportions.

It is of interest to recall that a similar though less definite activity is seen at dawn and sunset in other animals. The most outstanding is perhaps that of the howler monkeys of South America, which delight to sit in the tree-tops and make the welkin ring at sunrise and sundown. In them it is a matter of letting other groups of howler monkeys know which territory is occupied, before the daily foraging begins. The same is probably true for birds.

SINGING MOUSE

I was once called to a neighbour's house to release a bird trapped in the kitchen. According to reports it had been heard singing for some time, but nobody had been able to find it. A brief search of the kitchen showed that no bird could be seen. Yet the soft, sweet notes could still be heard.

Then, using the ears rather than the eyes for searching, we eventually found that the sound came from under the sink, where a mouse was perched in the angle in the waste pipe. In spite of our presence, it continued singing for several more minutes, before dropping nimbly to the floor and disappearing through a hole.

There have been many reports of singing mice from time to time. But why the mice sing is still something of a mystery: is it for pleasure, or from a feeling of contentment? It has also been suggested that they are indulging in a love-song.

It is, however, significant that the descriptions of the song vary a good deal, from a few low, soft notes to a song sustained for 10 minutes or more and resembling, as with my mouse, the song of a canary.

Postmortems have shown that singing mice are always males, and always there is an inflammation or other injury in the larynx.

ECHO-SOUNDING IN THE DARK

It is less than 40 years since the long-standing mystery of how bats find their way in the dark was finally settled. They use echo-location, or sonar. That is, they send out ultrasonic squeaks while they are flying and listen for the echoes of these reflected back from solid objects. These objects include the insects they are hunting.

Moreover, during those 40 years it has been found, unexpectedly, that porpoises, shrews, mice and others also use ultrasonics for echo-location. Baby mice in distress use ultrasonics but not for echo-location. They call for mother, in voices too high for our ears and probably also for those of any listening cat.

In fact, the discovery about the bats opened the floodgates to all manner of new information, including the fact that two kinds of birds are known to use echo-location, though not ultrasonics. They are the cave swiftlets of south-east Asia and the oil-birds of South America that roost and nest in caves in absolute darkness. They make clicks audible to our ears to find their nests in the dark but stop making them when they leave the caves.

Another form of echo-location, but one which cannot strictly be called sonar, is in even wider use. It involves the use of sounds not deliberately made for this purpose, just as blind people can be guided by the echoes of their own footsteps bouncing off solid objects. Rats, for example, are constantly making small noises, including grinding their teeth to keep them sharp. Experimental rats, blindfolded, can readily find their way around obstacles; but not if their ears are blocked so that they cannot hear their own teeth gnashing.

BIRD VOICES IN THE NIGHT

Several reports have reached me of cuckoos calling right up to nightfall or even after dark. Although not unknown before, it was possibly more frequently heard in 1952. There have also been a few reports of a cuckoo heard calling vigorously from midnight through the small hours.

The use of song or a call-note is primarily to advertise the boundaries of a territory occupied by a pair of birds. Cuckoos are no exceptions, for although they neither build a nest nor incubate the female tends to keep to one locality; the male also, with perhaps a slightly greater tendency to wander.

A song thrush: the purpose of bird-song is usually utilitarian, yet is it not possible that a thrush may sometimes sing for the pure joy of it?

Song is also used for several other essential purposes during the breeding season but this does not rule out the possibility that a bird may sing or call just for the love of it. Blackbirds and robins sometimes sing after dark, so why not cuckoos?

Birds often call out in the night because they are disturbed, either when an enemy is on the prowl or for some even less obvious reason. Even a strange noise will sometimes start them off. According to circumstances, the disturbance is greeted with a short burst of song, more commonly with the alarm note. Cuckoos have no special alarm call, so we cannot tell, in the middle of the night, whether they are calling in apprehension or from force of habit.

Moreover, while most birds disturbed at night may pass unnoticed except at close range, the stentorian notes of the cuckoo compel even more attention when the rest of the world is quiet.

NOISY NIGHTJARS

Nightjars are always described in the books as crepuscular in habits, which means they are active only at twilight and no more. Then why are they called nightjars?

During the past few weeks, whenever I have gone into the woods after sundown to see them, the tawny owls have been out some minutes before the nightjars. But these owls are universally accepted as nocturnal—active throughout the night.

It was natural to assume, therefore, that nightjars, after greeting the approach of night with their long churring trills, followed this with a short spell of intensive feeding and retired to their resting places in the bracken or the heather until daybreak.

A fortnight ago I woke up by chance at 2.45 am. Going to the window, as I usually do on such occasions, I leaned out to listen to the night sounds. Almost at once I heard a cock crow, and the call was taken up by half a dozen others. As the clock struck three the first nightjars struck up, and they continued trilling almost until the first songs of the day birds began.

Since then I have heard them at various times between twilight and midnight, at 1 o'clock and round about 2 o'clock in the morning. It may be true that nightjars are crepuscular in their habits.

27

Nightjar

Nevertheless, the nightjars living in my neighbourhood are, throughout the night, more evident to the ear than the numerically stronger, and truly nocturnal, tawny owls.

WHEN THE VIXEN SCREAMS

On three occasions in the past two weeks a fox has passed through a nearby meadow repeatedly screaming. It has always been the custom to speak of 'hearing a vixen scream'; in recent years has come evidence that the dog-fox also will scream, although in my experience the vixen does so more frequently.

Whether 'scream' is the best word to use must remain a matter of opinion. Moreover, the red fox uses so many vocalisations we cannot always be sure which cry is being described when someone claims to have 'heard the vixen scream'.

This one is a weird, almost unearthly cry, part scream, part wail, yet a thin sound, almost ghostly. It seems to be unrelated to any seasonal function since it may be heard in any month of the year, perhaps repeated on several days running and then not heard again for weeks or months.

Another baffling feature is that any fox in the immediate neighbourhood does not respond and seems almost to ignore the scream. So it appears to be not part of a language or exchange of signals.

I have identified this same spooky sound in a dog-fox whose mate has been recently killed and in a vixen bereft of her cubs. More often it has no obvious association with grief, discomfort or disquiet on the animal's part. It seems to have nothing to do with loneliness, the breeding season or hunger.

So the reason for this typical countryside sound is still a mystery.

THE BAFFLING WOODPECKER

It was fairly quiet as we passed through the copse. Then the alarm note was sounded. Recalling the alarm call of a blackbird, its

Great spotted woodpecker and blue tit

metallic quality and its frequency left no doubt that here was a greater spotted woodpecker.

The call went on and on, but no bird could be seen nor could we say precisely from which point in the trees the voice came. In the end we approached more closely and finally stopped under one particular tree. First the call would be muffled, seeming to come from low down inside the trunk. Then it would be high up but still inside. Or it might appear to come from outside the tree or even from a neighbouring tree, and then back inside the tree again. At last we found the 2-inch hole, the opening to the woodpecker's nest, 8 feet up the trunk.

This was no ventriloquist, but an agitated bird that moved up and down inside its nesting chamber, a hollowed cavity inside the trunk, and periodically thrust its head half out or right out of the entrance hole.

As the bird moved, so the volume and direction of the voice seemed to change and the voice itself to run up and down inside or outside the tree or from point to point over the clump of trees. The human ear is a poor thing for judging direction, but the chances are that even the efficient ears of the woodpecker's natural enemies would be a little baffled.

OWLS' AUTUMN CHORUS

The tawny owl's full song period begins in mid-January and continues until the beginning of June. This covers the breeding season, with a certain amount of overlap before and beyond it. During this time, although the owls use a variety of other calls, those most noticeable consist of duets, the male hooting and the female responding with a sharp 'kewick'—the owls' equivalent of the more musical phrases used by other birds.

From early June to late September the owls are relatively quiet, but during the tail-end of September and for most of October there is a mild revival. The chorus may not be as loud or as sustained as that associated with the breeding season, but it does represent a minor autumnal peak.

So we find people are apt to comment in autumn on the noise the owls make at night. The offender is bound to be the tawny owl, for it is fairly commonly distributed over the country—not so common in the west and rare in the extreme north. Its natural haunts are woods, copses and parklands. Inevitably it has spread into gardens with well-grown trees, many of which are now included within densely populated suburbs.

A NIGHTINGALE SANG

This was one of the rare moments. Returning from an early

morning walk through the woods we reached a clearing just as the sun had risen above the brow of a low hill. Its golden light flooded the air and seemed to accentuate the radiant greenness of the fresh vegetation. And a nightingale started to sing.

I have listened to nightingales at all times of night, and at most times during the day. Sometimes their notes are only half-hearted, and sometimes their persistent singing throughout the night can be tedious, if not positively nerve-racking, as it intrudes on one's sleep. I have often heard discussions on whether the song of the nightingale is as good as it is made out to be, preferences being expressed for robin, blackbird and thrush. Each of these at its best can rival, if not excel, the nightingale—except when that bird is at its best, as it was on this particular morning.

There were few preliminaries before it reached the full flood of song. Then every note and phrase in its repertoire was used, with hardly a pause, for a full ten minutes, after which it stopped abruptly. But during that time the thicket of young saplings and brambles seemed to be filled with a tangible, if invisible, mass of melody.

The nightingale sang as I have never heard one sing before; and never again shall I doubt which is our loveliest songster.

MAKING A SONG AND DANCE ABOUT IT

Grasshoppers were filling the air with music 200 million years ago, when birds were still little better than feathered lizards. In common with other primitive insects, their life-history shows an incomplete metamorphosis. There is no larva and pupa, each succeeding stage in the life-history differing only slightly from that preceding it. My picture (*over*) shows three stages of development.

There are two kinds: the short-horned grasshoppers, with short antennae, related to locusts, that lay their eggs in the ground, and the bush crickets or long-horned grasshoppers that usually lay their eggs in neat rows in the stems of plants. In both, the female has a sabre-shaped ovipositor for inserting the eggs.

On hatching, the grasshopper has only the smallest sign of wings but is otherwise similar to the adult except in size. Further growth

31

Three stages in the growth of a grasshopper

is by a series of moults. At each of these the body becomes longer and the wings larger.

We see little of the insects during these early stages. The eggs are laid at the end of summer, remain dormant through the winter and hatch in spring. The young grasshoppers keep out of the way under stones and logs during the vulnerable, almost wingless, period.

In July, they show themselves, or better, make themselves heard. Like other living things, they make more music when the sun is shining, and it is a true love-song. Short-horned grasshoppers make it by rubbing the hind legs on the wings, bush-crickets rub the wings together. But only the males make music.

CHAPTER 3

MAKING LOVE

THE MARCH HARE

It has been my ambition for many years actually to witness the performance of the mad March hares. So far I have been either too late or too early, or have been in the wrong place. In fact, the extraordinary thing about this supposedly familiar annual event is the remarkably small number who seem to see it.

Those that do so describe the bizarre antics of the males, or jacks, during the rut. They are said to gather in groups, often coming to the same spot several days running, to indulge in wild antics. They leap on stiff legs, bucking and kicking, and pairs or groups indulge in stand-up boxing matches.

This is not the only aspect of the lives of hares about which we are generally ill-informed. For example, little precise information exists on how far they travel, except that each hare is believed to have a home range of 1–2 miles diameter. Even if individual hares do not travel farther than this, the species itself is wide-ranging, for although we call it the European hare it is found not only throughout Europe but also most of Asia and the greater part of Africa.

Is the familiar hare a great traveller? There are reports of hares homing from up to 290 miles from the original point of capture.

HERON'S ERRATIC WOOING

It sounded as if a dog was yelping in distress down in the plantation, but the noise came from the tree-tops. Fifty or sixty feet from the ground was a mass of sticks, about two feet across and the same deep, and on it was a heron.

Away to the east another was flapping its way in, and as it drew near the call of the bird on the nest changed from the dog-like cries to a high-pitched single note.

At first it looked as though the newcomer would settle, but instead it circled the tree-top and, still in a glide, disappeared. Meanwhile the male, the one on the nest, had thrown aside the familiar statuesque pose of the heron, and with its long neck outstretched, it lowered its head into the nest, clapping its beak like a pair of castanets. The sound came thinly through the trees, as of two pieces of light board being clapped together. The heron's wooing is vigorous and noisy, remarkable for its wide range of sounds. But it is hardly musical.

The hen responds to the cacophony with a cautious approach, flying round and away, later returning, perhaps to perch on a nearby branch. So she will come gradually closer until finally she enters the nest. She knows instinctively that if she comes in too eagerly her wooer, in jealousy for the nest, is likely to mistake her for a rival and attack.

MALLARD'S AUTUMNAL IDYLL

Every autumn mallard get up to strange antics. The drakes particularly seem to behave with solemn foolishness, gathering in groups and swimming round with their heads sunk into their shoulders. Then suddenly each will dip the tip of the bill into the water, raise his body high, and rapidly pass the bill up the breast, throwing up a slight shower of water.

After this there may be dipping the breast and raising the tail, two or three times in quick succession, or a quick head-up tail-up movement so that the bird appears foreshortened. Finally, the drakes swim about with necks stretched low over the water.

This communal dance, this ceremonial, is repeated again and again, and like all dances follows a set pattern. The ducks persuade the drakes to it, and then go to one side, interested spectators. After a while, however, one of them comes forward, picks on a drake

Mallard drake displaying

and leads him off, or rather she invites him and then follows—and makes sure no other drake does. Away on their own, the pair continue the ceremonial display, but now it is a private affair, the pair cementing their partnership by a variety of formal actions.

Preening and drinking are examples. The drake will drink, then lift one wing slightly and reach behind it with his bill, as if to preen; but instead he runs the bill over the heel of the wing, making a sound as of a wooden rattle. This is called mock-preening. Sometimes the drinking, too, is only a pretence. Neither is accidental, for immediately afterwards the duck will repeat the actions.

Sometimes the drake will thrust himself forwards and upwards until finally he stands high out of the water. He may perform the grunt-whistle—thrusting his bill vertically in the water, then hoisting himself high again. The action stretches the windpipe, producing a whistle and a grunt. Or he may swim with head up and tail up, or bob up and down. All actions are as set as the movements in ballroom dancing.

Really we may call this keeping company, for as autumn runs into winter, the outward show becomes less, but there is obviously a strong bond between the birds, and neither lets the other out of sight.

Nesting does not take place until March, or thereabouts, but in the meantime there is work for the pair to do, if only to choose the site of the nest and get used to foraging the territory around it.

GOT A LIGHT, MATE?

For most of us the interest in glow-worms is romantic rather than utilitarian. This has always been so, except for a somewhat doubtful recipe centuries ago for making a lamp with the aid of a glow-worm and some quicksilver. One practical-minded reader, however, wonders how far this beetle could be used in the garden against snails.

It is common knowledge that the wingless female glow-worm carries a light which glows white with a greenish tinge. She has to turn the end of her abdomen sideways to show the luminous patch on the underside of the last two segments.

It is usually assumed that she shows this light to disclose her whereabouts to the male. This may or not be true. As in all instances of animals having light-organs we are compelled to guess the purpose they serve.

The male glow-worm also is luminous, but to a lesser degree, and the little light he can put forth is largely hidden by his long wings. Since he alone has a wide freedom of movement, his luminosity seems to serve little purpose unless to reassure his lady-love that he is on his way.

The adult insects, which show their lights in summer, eat little or nothing at all. It is the larvae that prey on snails, but they would have to be unusually numerous to make any marked clearance in a garden.

RING ROUND THE DOE

Roe deer used to inhabit the forests all over Britain. Now they are plentiful in many parts of Scotland and northern England, where they are native. They have appeared again in other parts of

36

England, especially in the southern counties, as a result of being re-introduced.

In early summer their newly-grown coats are fox-red and they are apt to be mistaken for red deer when seen at a distance by anyone unfamiliar with them. But a roe is only half the size of a red deer.

July sees the start of the roe rutting season and of the roe rings. As with so many other things about animals, it is not easy to account for these. We know that the more or less circular tracks trodden in the grass are made during courtship, and that typically a ring has a tree or bush as centre. Sometimes the shape is a figure of eight.

A red-deer stag forms a harem of hinds later in the year, and may then wear a ring around the harem with his hoofs as he continually makes the circuit of his wives to keep other stags away.

The roebuck may have some of this possessiveness, which would explain the roe rings. All we can say is that as the pair move round the ring the buck is on the outside, and that he will shoulder the doe or head her off if she shows signs of leaving the ring.

NO GENTLEMAN

Humble but familiar, at most 4in long, the three-spined stickle-back would probably have remained unknown to most people if it did not provide such a suitable target for small boys becoming in-itiated into the gentle art of angling.

The fish has another claim to fame: it has an elaborate courtship, which under the cold eye of scientific scrutiny can be broken down to nothing more romantic than a series of actions and re-actions between the sexes. Thus the male builds a nest and estab-lishes a territory. He attacks any fish coming within that territory, but if the intruder happens to be a female with a swollen belly he recognises her as being ready to lay, breaks off his attack and does a zig-zag dance in front of her.

She recognises him by this dance and also by the red throat which is his courtship dress, assumed in spring as part of this cold-blooded romance. By the zig-zag dance the male leads the female to his nest and by nodding at the entrance indicates to her what the next step should be. She enters the nest and he prods her in the flank, stimulating her to lay. Once she has done this he drives her away without mercy.

Another claim to fame is that the stickleback is voracious out of all proportion to its size. An eighteenth-century naturalist has bequeathed us an account of how one of his pet sticklebacks ate 74 young dace, each a quarter-inch long, and two days later ate a further 62.

LOOKS NOT ALL

A Surrey reader told me a nice story. Some months before a dishevelled robin began visiting the back (north) of his house. Its upper beak was half the length of its lower, so he fed it on soft food. In the rain, the bird would appear on the window-ledge wet and bedraggled, being unable to preen its feathers.

Another robin already tenanted the territory on the east side of the house and the two birds would put up a spirited defence if either dared poach on the other's domain. But recently Broken-Beak came for his food accompanied by the other bird. He was clearly introducing his fiancée. My correspondent thinks, however, that mating will be doubtful owing to Broken-Beak's unprepossessing appearance.

I had hoped that the reader would keep me informed on the course of events. It is still an article of faith with zoologists that appearance matters; that the cock with the finest plumage gets the pick of the hens, as Darwin taught us. I must confess to a scepticism in this. With birds there is much to support the argument that spirit counts more than plumage.

Another reader living in Devon, told me how a cock blackbird began to visit his garden in April 1952. The bird's feathers stood on end, and his upper beak was broken and a quarter of an inch shorter than his lower. By autumn he had acquired a mate and looked very sleek in his new feathers. Every year since Old Ragbag has had a mate and has nested in the same tree in the garden. He has never looked quite as untidy as he did in that first April.

As I said before, it is personality that counts—beak or half-beak.

In November the nesting season for rooks is still some months ahead but the preliminaries are already beginning. The large flocks in the air show energetic aerobatics due less to gusty winds than to high spirits in the birds themselves.

The rooks also congregate on the ground, not spread out for feeding, but rather in assemblies. Even so they do feed, but these gatherings, whether in the air or on the ground, have much in them to suggest social occasions. It is the time when partners are being chosen. Every now and then we can see a cock paying court to a hen. As with our own race, some couples get engaged quickly, others take some time to make up their minds.

There is the story of the jackdaw that tried to emulate the peacock with borrowed plumes. His cousin the rook does pretty well with his own plumage. A male rook displaying to his chosen hen is a magnificent sight. He may not have the colours of the peacock, but his strutting is every bit as vigorous, as stately and as statuesque.

He holds his head high, droops his wings and paces towards her with his tail feathers held almost vertically in a surprisingly handsome fan. Indeed, to watch the performance is to realise that, if a rook had the same long plumes and the same colours, his courtship display would rival that of the peacock.

TURTLE DOVES WOOING

At the end of April or in early May the turtle doves return from their winter quarters in tropical Africa. They nest mainly in the south and east of England and in the Midlands, and although they may breed in other parts of the country, they are there mostly wanderers.

That the turtle dove has a family tie with our familiar wood pigeon is obvious. The two have the same general build and habits, but the dove is the smaller, the more elegant and the more colourful. Even at a distance, when the smaller details of the plumage cannot be seen, there is an impression of pinks and fawns, and the white edge to the tail stands out clearly.

Turtle doves' courtship display

In June the call of the dove is apt to be somewhat quicker and more vibrant than usual, for courting is in full swing. It seems deceptively near as, on a tree perhaps a field away, the cock bobs and bows in rapid time to the hen, puffing out his throat and raising the feathers on his neck so that his head looks grotesquely swollen.

He advances towards her—and she flies on to another branch. He follows and again he bows and curtsies, and again he tries his luck. So, from branch to branch she leads him on. He may show off as much as he pleases, but it is the hen that sets the pace. And she does so by the oldest of stratagems—an air of indifference.

COURTING FOX-TROT

Foxes keep themselves very much to themselves. They are adepts at moving through cover and will perform a vanishing trick the moment they hear, wind or sight a human being.

But there are times, in spite of this ingrained cautiousness, when they may throw discretion to the winds and turn playful.

February, though it varies with the part of the country, is the courting season, and when the dog and the vixen are paired off they begin a courtship play. This can at times be vigorous. There is much chasing, and one of the distinctive features is that repeatedly, during a rough-and-tumble, the dog throws his hindquarters at the

40

vixen's shoulders and wraps his brush over her. They hold this position for a brief spell before breaking into the rough-and-tumble once again.

At all times, the play is carried out with wide-open mouths, to the accompaniment of staccato, almost querulous sounds, as if a fight were about to break out. That is only their way of showing excitement, and at intervals the two will drop into a gentler mood, as the dog grooms the vixen's face or behind her ears. Every now and then the play reaches a climax as the two come face to face. With open mouths almost interlocked, both rear on their hind-legs, place the forepaws on each other's shoulders and perform a rather irregular pas-de-deux.

ACROBATIC COURTSHIP

In February the flocks of lapwings have not yet broken up. On sunny days we see them in the sky, two to three hundred or more, moving aimlessly around, recognisable even from a distance by their wobbly flight. Yet within these flocks the formation of pairs has already started and before long the pairs will begin to separate for the breeding season, each pair settling into its chosen territory.

Building a nest occupies little time, a scrape in the surface of the earth being sufficient for the inconspicuous, dappled eggs. These will not appear until the end of March or the beginning of April. Before then, we shall see the acrobatics of the male—a striking feature of the courtship.

Low over the fields or marshland he will fly, with a strange leaping and bounding flight, uttering his mournful cry of *peewit*. Every so often he will drum with his wings, a quaint beat which will sound across the fields like a powerful bird flying hard against the winds.

Not content with this, and as if out of sheer exuberance of spirits, he will occasionally interrupt his clumsy flight to execute a tumble, twisting and turning, as if mortally wounded. Then he will fly up again, and again we hear the mournful cry and the boastful drumming of the wings. A wheel and a sweep and he is down to earth again, to pause for a while before resuming his antics.

Male lapwing drumming

COURTING HOVER-FLIES

Hover-flies are common enough, so we seldom spare them a thought. Occasionally we may stop to wonder, with the larger kinds, whether we have to deal with a bee or a wasp. But there can be no doubt once we see them hover.

This is a masterly performance, as they hang suspended in the air, swaying slightly, backwards and forwards, then flying forwards, sideways or backwards, only to fetch up again in a hover.

Such skilful flyers are not content with an ordinary courtship. The pair hover opposite each other, as if each were suspended by an invisible thread. Then at intervals and quite regularly they fly towards each other, bang their heads together and swing apart again. While this is going on the two sink slowly to earth, and all the time the male keeps up a shrill hum. Also, the hum of the female's wings catches the attention of the male and in response he raises the pitch of his own humming. The sound of their wings is therefore a simple form of conversation, which can, in fact, lead to difficulties.

Hover-flies have been seen to charge other insects and knock them off balance. They will even ram the very much larger bumble-bees. What looks here like malicious intent is merely a case of the male hover-fly mistaking the hum of another insect for the call of love.

CHAPTER 4

SELF-HELP
AND SAMARITANS

PIONEER THRUSH

A Berkshire reader describes sitting at her window one day watching a thrush making vain efforts to smash a snail shell on the brick edging of a rose-bed. After some time the thrush dropped the shell, paused for a moment, then hopped across the bed to return with a piece of glass in its beak. It then attacked the shell with the glass, finally cracking it and eating the snail.

Intelligence, it was once suggested to me, is the ability to recognise a problem, assess its possible solutions, select the most likely solution and act upon this quickly. Did this thrush show intelligence?

There are remarkably few animals, outside the apes and monkeys, known to use implements. Among birds, one species of Darwin's finch uses a piece of stick held in the beak to probe insects out of bark, and one of the bowerbirds makes a brush for applying coloured juices to its bower. In 1967 it was discovered that the Egyptian vulture may throw a stone at a large egg to break it open. There are only about half-a-dozen others I can think of in the whole animal kingdom.

CROWS' COMPASSION

In a Note written in October 1955, I remarked that it is not fashionable nowadays to credit any animal with emotion, and an Essex reader has taken me up on this. My remark, intended to be mildly satirical, would have been nearer the truth had I said that

A carrion crow

there was a disbelief in animals having the finer emotions. Nobody doubts that animals experience fear, but if I were to state that a bird may sometimes sing for the joy of it some people would disagree with me.

An ex-gamekeeper friend told me how a fellow keeper shot at and winged a crow. The bird fluttered to the ground, and soon two crows landed, one either side of it, and supported it with their outstretched wings. My friend and his fellow keeper could not stay to see the end of the episode because more and more crows assembled overhead and began to behave in a menacing fashion. I know my friend to be a good observer and not given to romancing.

But why do I have to support his story thus, with a testimony to his personal character? Not because anything he said is physically impossible but because we do not normally credit crows with the ability to show compassion.

SMALL BRAIN WITH A LARGE MEMORY

The solitary bees are building in the sandy bank again. We watch them coming and going and wonder how each can find its own nest, for the scores of circular holes in the sloping bank all look alike and many lie close together.

They find their way home, of course, just as we do, by sight and memory. It is easy enough to put this to the test by removing the scraps, the pieces of twig, pebble and the rest that lie around a particular entrance. If these are carefully arranged in the same pattern elsewhere the bee can be made to enter the wrong hole.

These smaller landmarks come into play at the end of the journey only. In its return from a nectar-gathering trip on the nearby sallows the insect is guided first by a gross assessment of the larger features of the landscape.

Then the region of the nesting burrow is recognised from the smaller landmarks, a log, a large stone here, a clump of grass there, and so on. Remove these and a returning bee is lost, and an outgoing bee must make laborious reconnoitring flights to get its new bearings.

There is, therefore, an equivalent of our street names and house numbers, and a visual memory similar to our own. The wonder is that one small head can carry all a bee knows: for its so-called brain, composed of a few nerve-cells only, is smaller than a pin's head.

RED ANTS AND THE CATERPILLAR

For a long time naturalists were puzzled to know why the caterpillars of the large blue butterfly were never found. There were plenty of adult butterflies but apparently no caterpillars. Then the reason was discovered. In June the female large blues lay their eggs on wild thyme and 10 days later the caterpillars emerge. For 3 weeks they feed on the flowers of the thyme, and they also devour the smaller larvae of their own kind, or those that are in the process of moulting—an act of cannibalism.

After the third moult the remaining caterpillars drop to the ground and wander aimlessly until discovered by a red ant, which may be one of two species. Instinctively the ant recognises that the caterpillar exudes drops of a honey-like substance, which the ant enjoys. The ant strokes the gland exuding this, sips the syrup, then picks up the caterpillar in its jaws and carries it back to its nest. (Ants' nest is something of a misnomer. We ought to find a better word for this kind of communal shelter.)

In the nest the caterpillar feeds on the ant larvae and grows rapidly, the ants frequently stroking it to milk it for its syrup. During the winter the caterpillar hibernates, completing its growth the following spring, when it pupates. Three weeks after this, the adult emerges and crawls off.

The butterfly seems to have become parasitic on ants and is apparently unable to survive without them. More remarkable, it has been found in recent years that the ants become so addicted to the sweet substance given off by the caterpillars that once they have carried one to their nest they not only allow it to feed on their own larvae but they also neglect these larvae. In time the colony of ants harbouring the caterpillars tends to degenerate.

This curious perversion is regarded as being due to an aberrant instinct—an instinct that has developed along the wrong lines.

The scene was a back porch of a cottage in Devon. A pair of blue tits had built a nest between the sloping slated main roof and the flat portion of the overhanging porch-roof. My correspondent says she had often tried to see fledgling tits leaving the nest, but without success. It is one of the frustrating things about these small birds; they are so indifferent to the human presence that we can watch them building the nest and everything else, but the young leave so suddenly that before we know it the whole family has gone.

On this occasion, my correspondent was at a window close to the nest when the chicks appeared, with the parent bird apparently coaxing them out. The first emerged and followed the hen to a railing immediately below the nest. From there it was induced to follow her to a nearby holly hedge. Two other chicks did the same, but one remained and seemed scared of leaving the nest.

At last the mother managed to coax it to the railing, but it refused to follow her to the holly hedge. The hen flew away, and reappeared shortly after with a large grub held ostentatiously in the beak. This she offered to the 'reluctant debutante', which tried to grab it, but the mother drew it away and flew off a short distance. Then she returned and repeated the performance. At the third attempt the chick took to its wings and followed her to the hedge.

HELPFUL CHICKS

Moorhens usually bring off their first broods in June. The chicks are like farmyard chicks in the length of their legs, their large feet and the small wings, but there the resemblance ends. The moorhen chick is covered with a black down, except for the patch of blue skin over each eye, and the bill is red with a yellow tip.

There is another, more striking difference. The moorhen chick is more precocious. The farmyard hen helps her chick by indicating food with her beak, leaving the chick to pick it up. The moorhen parents proffer food with the beak and the chicks run or swim for it.

Moorhen chick

In other ways, too, the moorhen chick shows a greater alertness and ability to look after itself. Later there may be a second brood, when the chicks just hatched will be half-grown. Then will come something rare in the animal kingdom: the first generation of youngsters may help to feed and care for the second brood.

All the time they are sitting the parent moorhens are adding to the nest. Should the level of the water below it rise, the addition of new material is hastened to keep the floor of the nest above the flood waters. When this happens the parents collect the materials and may pass them to the half-grown youngsters who build them into the nest. There can be few other instances of young birds helping in nest-building.

DELAYED INSTINCT

We talk about house martins teaching their young to fly. It would be truer to say they encourage them. When the young martin first leaves the nest it may cling to the side of it for a while. Adult martins from all around congregate in a flying circus and one by one they zoom over the clinging youngster or land beside it for a moment before taking off again. They seem to be showing it what to do.

All this would not be necessary if the young martin could fly at first attempt. Instead there seems sometimes to be a lapse of time between leaving the nest and becoming fully airborne. Occasionally a young bird will flutter and glide, landing on a lower roof near the nest. There also the ceremony of adults zooming over it or landing beside it is enacted, until finally, after a fair interval, the youngster takes strongly to the wing. On 19 August one year, two young house martins landed on the ground below the nest. They remained there for fully half an hour, seeming content, stretching a wing from time to time, while half a dozen adults appeared to be encouraging them to take off.

A reader told me of finding a young martin on the ground. He picked it up and tried to launch it, but to no purpose. He took it home and, for safety, placed it in a shed, where it seemed to be content to do nothing. Twenty-four hours later, after having several times failed to induce it to fly, he held it in his hand. This time the bird stretched its wings once or twice and flew off strongly.

WINGS OF RESCUE

This story sounds incredible, but as soon as I was told it I visited the site and questioned the two witnesses, and find no adequate reason to doubt what they told me.

Two men working in a sandpit in Surrey had taken a day-to-day interest in the colony of sand martins nesting in the bank. One June day one of them found a nestling on the ground, chirruping for dear life. They tried to get up to replace it, but the nests were out of reach, so the nestling was replaced on the ground where it had been found.

It was their midday break, and they sat watching in a more or less desultory way. About three-quarters of an hour later they saw two sand martins fly down and bear the nestling up to the cliff-face and go into one of the nesting holes. The best they could say was that each adult appeared to grasp a wing in its beak, and that in flying up the right-hand adult seemed to be beating only the right wing, the left-hand adult the left wing.

Even if essential details are missing the story is worth putting on record. Accounts of adult birds rescuing helpless young ones are few and far between. Always they sound incredible, but there are enough of them to indicate the possibility that rescues do sometimes take place.

QUICK DECISION

Can animals think? We are often assured that they cannot. They may not use thought to our high standard, but have they at least the beginnings of it? Let us take the case of the squirrel about to cross a road.

It ran down a tree and was part of the way across the road as the wheels of a car drew near it. With lightning speed it braked, turned completely round, returned and shinned up the tree. The whole action can only be described as split-second.

We may say that had the squirrel thought, it would have continued its journey across the road and up a tree the other side—except that there were no convenient trees on that side. In the face

Grey squirrel

of rapidly approaching danger there was no panic nor faltering, but to all appearances a split-second decision to reverse all previous intentions and return to a point known as safe.

We may say it was an automatic reaction. An automatic reaction is stereotyped, like blinking the eyes when a solid body comes towards them. But in that no provision is made for alternative action. Here the squirrel had several courses open to it, to go on, to stop, or to reverse. It had a choice. A choice means a decision must be taken. Merely because the decision is rapid need not mean it is automatic.

FOXES WOOL-GATHERING

Every now and then the story of the fox and its fleas crops up. I have been asked whether there is any truth in it.

The story, briefly, is that someone claims to have seen a fox gather wool, then go down to the river, wade in and submerge until only the tip of its snout and the wool are above water. Then the fox surfaces and lets go of the wool, which floats away with the fleas from the fox's coat.

I first became interested in this story about 26 years ago, and since then have accepted any opportunity to probe it. Many people claim to have seen it happen, and a half-dozen have, at various times, given me their unsolicited testimony to having watched it.

In pursuit of evidence I have talked to shepherds, several of whom have told me emphatically of having seen foxes collect wool from hurdles or barbed-wire fences and trot off with a wad of it in the mouth. But they were unable to carry the story further than this.

All this amounts to no more than an unsure foundation, but I was impressed by a story volunteered by a young man in his late teens. He had, apparently, never heard about the fox and its fleas, but he described his own dog collecting wool and submerging in this same manner.

GRASS SNAKE SHAMS DEATH

Grass snakes usually make off rapidly when disturbed. This one was sunning itself and seemed in no hurry to move. So I picked it up. As I lifted it by the neck its mouth lolled open. The lower jaw hung loosely to one side, the tongue hung limp. It was a very good performance.

I do not remember seeing a dead snake that had not been battered to death, so whether one that dies a natural death really looks like this is hard to say. But there was no question that this one was putting on an act. It looked dead and the muscles of its body were rigid as if in rigor mortis. Then the tail moved slightly and gave the game away.

You could move the snake about as much as you liked; it still returned to this deathlike pose. What is more, you could move the head into a normal position so that the mouth closed and resumed a more life-like appearance. But if you now put the snake down on the grass it remained thus for a brief moment; then, with an abrupt movement, turned its head over so that the lower jaw was uppermost. At the same time the mouth lolled open, the lower jaw slipped awry and the tongue hung limp.

It was as though the snake were determined to appear dead no matter what you did about it.

A HERON USES BAIT

Should you miss goldfish from your garden pond it is not a foregone conclusion that the culprit is a heron. In 1964 a reader reported seeing a frog trying to swallow one of his goldfish. The circumstances suggested either the edible frog or the marsh frog. Both are imported from Europe and sometimes are liberated when the owners find their pets croak too loudly for their neighbours' peace of mind. Both these frogs take food underwater, which our native frog does not.

About the same time another reader described to me how he had seen a starling take a goldfish from his pond and fly with it to the lawn. A vigorous chase caused the bird to fly off and the goldfish was restored to its home.

Herons are always associated with taking fish but their diet list is a long one. One was seen to catch a bat and swallow it, and at the other extreme another was seen to steal bread from a duck in a London park. It had the sense to take the bread and soak it in the water before swallowing it.

Commonsense seems the last thing to expect from such a bird, but a relative of our heron went even better. This was an American green heron which found a piece of bread, carried it to water, dropped it in and seized the small fishes that came to nibble it. Moreover, it retrieved the bread each time it floated away, and later carried it to a fresh place where other fish were breaking surface.

SKIN OF ITS BEAK

A bird at roost normally stays put until dawn. In 1967 a Lincolnshire reader sent me a note describing what can happen when a bird is disturbed. During a gale he saw from his bedroom window a bird in a tall silver birch which, in the pale light of the moon, he took to be a song-thrush. The thrush, which was being blown about unmercifully by the wind, was using its wings and tail to full advantage in an effort to maintain its hold as the thin branch pitched and tossed.

Two hours later he looked out again and saw the bird still hang-

Song thrush

ing on grimly. This time he shone a powerful torch onto it and could see it was also holding on with its beak. It is particularly interesting that a thrush should have used its beak in this way. By dawn the bird was gone and there was no sign of it on the ground, so presumably it had weathered the storm.

No doubt this kind of thing happens more often than we suppose. A predator, such as a cat or an owl, disturbs the vegetation near the roost. The bird flies off, blind, and clings to any tree or bush on which it chances to land and there it stays, too frightened to move until dawn. In a gale the chances of this happening are much increased. A sudden gust of wind, or something blown through the trees, will frighten the bird as much as a living enemy would.

THROWING CRUMBS

A tame jackdaw, found straying, was brought to me 'for fear the cats might get it'. It was housed in an aviary, near the house, and proved lively and attractive. Each day it received its rations on an enamel plate set about a foot inside the aviary.

The next year a pair of blackbirds nested in a yew hedge beside the aviary. Later, the four young fledged and the parents were feeding them on the ground around the jackdaw's aviary.

Then we became aware that the jackdaw was passing food through the wire-netting of the aviary to the parent blackbirds, who gave it to their youngsters. This was continued. At times the jackdaw would run to the food plate, take up a beakful of food, go towards the wire-netting and flick its beak. Crumbs would fly in all directions, some through the wire mesh, and these were quickly swooped upon by the blackbirds. The whole thing may have started this way for all we know, and for the most part that was the pattern. Yet even this seemed purposeful. But every now and then, instead of flicking the beak the jackdaw would put it through the wire mesh and so hand the food to one or other of the blackbird parents.

The significant feature is that when one of the fledglings went to the wire, as they frequently did, and solicited food of the jackdaw by gaping and fluttering the wings, the jackdaw would repulse it. There seemed no reason to suppose therefore that the jackdaw was responding to a parental feeding instinct alone. Especially when it handed the food through the wire there was something deliberate about its actions. About a month after the performance ceased the hen blackbird appeared with two more fledglings and we saw again that the jackdaw was passing food to her.

Long before she had had her first brood she had been coming into the house to forage. She continued to come into the house as before, but now she brought her second lot of fledglings with her.

Altogether, therefore, she seemed to have established her own little Welfare State, and it appeared also that the jackdaw was entering more fully into the spirit of it, for this session there was a significant change in the way he dealt out his largesse. His daily rations consisted of brown bread moistened with milk, to which were added grapes, banana, raisins, sultanas, or hard-boiled egg. This session he picked out for the blackbird the delicacies, giving her bread only when the extras were finished.

More remarkable, almost every day, in the afternoon, we found the plate, bearing the remains of the day's rations, had been pushed half out of the aviary and the blackbird was helping herself. This could have been no accident because there was only one space around the base of the aviary through which the plate could be pushed.

RAT TO THE RESCUE OF RAT

The story was told me by two ladies who, disturbed by extensive signs of a rat in the garden, set a trap for it. A little while later they went into the garden and found a large rat caught in the jaws of the trap by the tail, near its base. The ensuing moments can be better imagined than described, as one went for a heavy stick to kill the rat and the other went to fetch a pail of water to drown it.

Both arrived back simultaneously a few moments later, just in time to see a second rat finish gnawing through the imprisoned tail. Then both rats were gone, the one leaving behind the greater part of its tail in the trap.

Common rat

There is no obvious reason for doubting the story, and, assuming it rests on accurate observation, it furnishes one of those rare examples of intelligent co-operation for which the student of animal behaviour is always looking. It may be such things are commonplace and well-known to those familiar with the ways of rats, but the event is significant in any case.

The behaviour of even the higher animals is largely automatic, or instinctive, but this still does not rule out the possibility that in emergency they are capable of rising to greater heights of intelligence. And because their everyday actions are largely selfish, it does not follow that compassion finds no place in their lives.

CHAPTER 5

TASTES IN COSMETICS

HOW BIRDS BECOME WATERPROOF

It seems absurd that water-birds should need to take regular baths, especially as the water flows over their backs like mercury over a polished table-top. What has been more puzzling is how the waterproof nature of the plumage is maintained.

It is doubtless due in part to the way in which the feathers overlap. A dead bird soon shows signs of becoming waterlogged, which may be because the muscular control of the feathers has been relaxed. The frequency with which birds preen themselves suggests, however, that having each feather in place is not only a matter of comfort but of practical use.

But preening is not merely to keep the feathers in place. The beak is first pressed against the preen-gland (in the so-called parson's nose), so that oil from the gland is automatically carried away and applied to the feathers as they are passed through the beak.

It has been suggested that another purpose of the preen-oil is anti-rachitic. When bathed in sunlight the oil is said to acquire a vitamin, and when the bird preens itself some of the now irradiated oil used in a previous preening finds its way into the mouth, and thence to the stomach.

The oil has another use, for it has been shown beyond a doubt that if the oil-gland of a duck does not function properly, the bird will sink. Perhaps even more interesting, it was found, certainly in ducklings, that the oil is formed only when the duck is able to eat insects or such food.

Birds are fastidious about keeping their feathers in condition. They show this by constant preening and by their addiction to bathing. But while cleanliness may be the result, it seems that there is more to it than this.

Preening and, more especially, bathing may be indulged from

Rook bathing

mere pleasure. Some birds will bathe in the rain and, when they do so, they seem to be seized with an ecstasy. Another peculiarity is that, contrary to what we might expect, they bathe more in cold weather than in hot. I have noticed this in wild birds, and it is certainly true of the two dozen aviary birds we have in the garden.

Another unexpected observation came in February 1958, during a period of fog. Every morning six gallons of water were needed to replenish the bath water in our dozen aviaries. Even if the water was not splashed out it became soiled in one way or another. During three days of fog the water remained unused and unsullied. There was no need to replenish it, even where, as in most of our aviaries, the bath water was also the drinking water.

When the fog was followed by a bright crisp morning, all the birds were bathing vigorously, rapidly emptying the containers of water. Once refilled, these were quickly emptied again. In fact, 20 gallons of water had to be carried in a morning to the aviaries by the time the washing frenzy was over, as against four or five gallons on a normal day.

MAGPIE FINDS A 'TOWEL'

Birds not only make a great fuss in taking their bath; they also make a great to-do about the preening.

The process includes a good deal of shaking, with the feathers fluffed out, and the bird pays much attention to the feathers with its beak. Both actions help to remove the water from the feathers. If we watch carefully, we can see water from the larger feathers dripping from the beak.

But what about feathers out of range of the beak—on the top of the head, for instance? A magpie, sufficiently absorbed in its bath not to notice that it was being overlooked, gave me the opportunity to see how these are dried.

Having finished splashing, it flew on to a small branch, and one of the first things it did there was to rub the back and sides of the head and neck on the surface of the branch. The action was vigorous but deliberate, with the nictitating membrane, or third eyelid, closed all the while, so that the eyes looked white. The appearance was as though the magpie were using the branch as a towel.

Magpie rubbing its head on a branch

60

All members of the crow family wipe the beak, vigorously and repeatedly, on a branch after bathing, but this is the first time I have seen one of them actually dry its head feathers.

SUN-BATHING THRUSH

'A few days ago,' wrote a reader in 1967, 'I was sitting just inside the French window of my room in the sun when a very tame song thrush hopped inside, as it often does, for a sultana. Instead of taking a second sultana offered, the thrush spread itself into the "sunning" posture, with wings and tail completely spread, and beak wide open, on the carpet inside the room and within a yard or less of me where I sat.

'It remained like this for some moments, then got up, ate a sultana and preened a few feathers, and repeated this performance several times—twice it seemed to be completely relaxed and the other two or three times partially so.

'I was utterly charmed at the confidence shown and wondered whether a bird has been known to come into a house to sun itself before. Also, did it mistake the plain green carpet for grass, or just choose the sunniest spot available at the moment?'

Sunbathing is common and widespread among animals of all kinds, and particularly noticeable in birds, but it has been little studied. No doubt there is a direct benefit from the sun's rays, but no less interesting is the way a bird appears to go into a trance, as it settles itself to bask.

I once watched a bird from a distance of about a foot and one of the first signs that sunbathing was about to commence was the faraway look that came into its eye, as if it were going to sleep with its eyes open.

BLACKBIRD GOES SUNBATHING

At a certain spot in the garden, just at the edge of the lawn near the base of a small cypress, a blackbird comes to sunbathe whenever the sunshine is warm. Presumably it is always the same bird. Certainly it is always the same spot.

If it is the habit of birds generally to have a favourite spot for this purpose, it is less obvious than with blackbirds. At all events, the regularity with which this one spot is used, when everywhere around is equally bathed in sunshine, suggests a ritual. And so does the way the blackbird approaches. I had seen it sunbathing many times; but not until recently had I seen the preliminary stages. There were five separate movements.

At the start, the bird was actually in shadow, standing with the wings directed obliquely downwards. It moved forward about a yard, stopped, and erected its crest. Then it moved forward about another yard, stopped and fluffed up the feathers of the back. Another yard forward, another halt, and, with the crest still up and the back feathers still fluffed, the wings were moved into a more fully drooping position.

Almost exactly another yard forward was taken, which brought the blackbird to its favoured spot. There it sank into the complete sunbathing attitude. The head leaned to one side, the beak came open, all the feathers of the body were fluffed out, the wings were fully spread and the tail completely fanned out.

FIRE WORSHIP

As October chills begin house fires are lighted and the time is on us when starlings, jackdaws and rooks may be seen bathing in the smoke. It is not a common sight, but where it does happen it seems to become a habit among some of the local birds to take turns at sitting on the rim of a chimney with wings spread and somewhat arched, the tail twisted to one side.

It looks as if the birds are enjoying the hot currents rising through the chimney, and that may be part of the story. It is, however, another form of the habit known as anting, in which a bird adopts this same pose while holding an ant in its bill and passing it under its wings.

Most four-footed beasts give fire a wide berth, but many birds seem to be irresistibly drawn to it. In times past they have been seen flying with burning embers in their bills, alighting on thatched roofs and setting them on fire. Pliny, the Roman naturalist, told of

Rook smoke bathing

this happening in Rome, and in the Great Chronicle of London for 1202 we read of 'fowles flaying in the Eyre berying in hire billes brennyng coles that brenned mony houses'.

The habit of picking up burning materials is not widespread, but one hears occasionally of a fire in the top of a tree or a nest smouldering, as if a bird had carried away a 'brennyng cole', perhaps a still burning cigarette-end or embers from a garden fire.

FIRE WORSHIPPERS

The heifers were scattered across the 30–40 acres of pastureland that bound us on two sides, peacefully grazing. The nearest was at least 100 yards away, the farthest more than double this. Then the bonfire of uncompostable rubbish was lighted.

In no time at all most of the heifers had stopped feeding and had wandered over to cluster where the smoke was billowing into the field. They stood in a tight group, as near the bonfire as the fence would allow, lifting their heads as if inhaling the smoke.

This is a commonplace event, wherever cattle and fire and smoke are in juxtaposition. When hedge-trimmings are burnt in a field where cows are pastured, the cattle will form a circle around the fire. When only embers and hot ashes are left some of them will straddle these, either enjoying the warmth or the wisps of smoke still rising from them.

Somebody once telephoned to tell me about being awakened by the barking of foxes just before dawn. A vixen and two cubs were in the garden for some time, making enough noise to wake everyone in the house. The highlight of their antics came when they found the still warm ashes of a garden fire. They scraped at these, raising clouds of fine ash and rolled in them in a seeming ecstasy. So I was able to add one more animal to the list I have been compiling ever since I saw my tame rook playing with fire.

We normally think of animals shunning fire, and suppose, if we think about it at all, that they dislike smoke. Yet one has only to recall how close a dog will lie to a fire in an open grate.

THRUSH ANTICS

Somebody once wrote to me about seeing a young thrush sitting upright with its tail flat behind it on the concrete pavement. It was pecking vigorously among the feathers under its wings and after it had flown away hordes of ants could be seen milling around the spot.

'Anting' is the name given to a piece of unusual behaviour among birds. Put briefly, a bird may be seen picking up ants one by one from the ground and apparently rubbing them on the feathers under the wing, first on one side and then on the other. It has long been a puzzle why birds should do this. The obvious explanation, and the one usually seized upon, is that it is a method of applying formic acid given out by the ants to the feathers to kill vermin.

The currently accepted version is that it is a process of feather maintenance, that in some way the formic acid tones up the plumage. One argument against both these theories is that, as I have established by very lengthy observations, a bird will ant three times on the left side for every once on the right side. If anting was a means of killing vermin we should expect to see birds three times more lousy on the right side than on the left.

Equally, if it were a method of feather maintenance birds that ant habitually, such as starlings, would have a glossier and presumably more efficient plumage on the left than on the right.

64

That, however, is a matter of argument. I am more concerned here with the way in which sunbathing by thrushes and blackbirds is sometimes mistaken for anting. In the latter the actions are noticeably rapid as the bird picks up ants and seemingly pushes them among the feathers on the undersides of the wings with vigour. The whole posture is more grotesque than in sunbathing, with an apparent air of excitement accompanied by contortions in which the bird will sometimes almost topple. The sure test is, however, to examine the spot afterwards to see if there are ants running around.

Anting is most frequently seen in August, although it is far from common even then. Ants are more numerous anyway because their populations have been building up during the summer, and there are also swarmings accompanying the nuptial flights of the queens. Another argument against anting being a form of feather maintenance is that at this time of year most birds are over the moult and their plumage is in first-class condition.

SUNBATHING ALREADY?

Lizards bask in the sun in Spring even if there is a cold wind, provided there is also clear sunshine. The habit is so pronounced and so much a ritual as to leave little doubt that the sun's rays are essential to the animal's wellbeing. Each lizard has its favourite place and, coming out from its hiding place, will make straight for this spot if the sun is shining. Then quite deliberately it takes up position, often on a stone or rock, and flattens its body to the ground to expose the maximum surface to the sun's rays.

Adders also bask in spring, and there is good reason to believe that the rays of the sun are essential for the female adder's ovaries to ripen.

Many animals bask, and it is not unreasonable to suspect that they all derive some physical benefit from it. We know that when a rabbit washes its ears it takes into the mouth natural oil from the skin that has been irradiated by the sun. This is a natural preventive of rickets. The preen-oil of birds, exposed to the sun and taken into the mouth during later preening, also has the same effect.

Lizard basking

Sun-worship has its practical side although the worshippers may be unaware of it. They bask first because they are drawn to do so by the sun's light, and secondly because the effect gives them pleasure. They enjoy sunbathing as much as we do.

OUT OF ITSELF

We found the adder's slough among the long damp grass. It was almost entire except for tears where the skin had come away from the jaws. My companion remarked, as he examined it, that he found some difficulty in taking off his socks let alone a whole skin, a joking remark that pinpointed a truth.

Casting a skin, even when done so neatly as this had been, is a trying affair. Two to three weeks before it happens the snake loses colour, its eyes look dim and it goes off its food. It also becomes irritable, liable to strike without cause or with little provocation. Then, just before the moult, its eyes clear again and the skin-colour returns. The snake rubs its chin on something solid to split the skin around the jaws. Then, if the reptile is healthy, it slips out, leaving the dry, horny covering behind, a kind of ghost of itself.

How the outer covering becomes detached from the underlying skin is not yet fully understood. Possibly blood flowing into the living skin carries enzymes that dissolve the inner layers of the outer covering so that the snake, for a while, lies as if in a closely fitting polythene bag, loose except around the margins of the jaw.

Snakes, unlike lizards, have no eyelids, each eye being covered with a transparent scale known as a spectacle. A sloughed skin is complete down to the pair of spectacles, which explains why the eye grows dim—while the enzymes are eating the spectacles from the surface of the eyes.

CHAPTER 6

PLAY OR NO PLAY

FUNLESS FAUNA

It may be that present conditions of agriculture, making the maximum use of land, have changed the habits of wild animals. It could even be that our network of roads, with its burden of noisy traffic, to which is added the chatter of the tractor and the roar of aeroplanes overhead, keeps them more under cover.

Whatever the cause a certain element of fun seems to be lacking today compared with the natural history of 50 or 100 years ago. Then it was not uncommon to find in magazines, and even scientific journals, entertaining accounts of grown-up weasels, stoats and foxes playing in the open in a completely absorbed manner.

For several years now I have kept a close watch without success for processions of stoats, a hundred or so strong, moving across the country in pairs. Nor have I seen the columns of mice or phalanxes of rats of which our forefathers wrote. I envy the observer who years ago reported seeing 100 or more shrews moving in single file each with a twig in its mouth.

Have animals lost a sense of fun, or are we more truthful in describing what we see?

PLAYFUL MALLARD

Mallard pairs are beginning to form up in October though the birds are still in flocks. They are getting engaged, so to speak, and occasionally a faint ghost of the springtime displays can be seen.

But there is one piece of display which has probably little to do with forthcoming domestic affairs. The drake seems to fly just under the surface of the water, making a great splash in the process, after which he flies up and lands again a few feet away. This is usually repeated several times. The orthodox explanation is that this is the escape reaction to the presence of a hawk, and that it is often carried out with no hawk near. This may be true. It is not easy to read a mallard's thoughts.

There was a time, however, when I had a pair of tame mallard in the garden. The best we could do for them in the way of a pond was a zinc bath and a pile of bricks beside it by which they could climb up and jump into the water. The drake spent a good deal of time climbing into the bath, swimming furiously under water, leaping out to land a few feet away, then waddling back and starting all over again. It may be an escape reaction; but this drake seemed to be doing it for the fun of it.

SQUIRREL TANDEM

Animals normally behave according to recognised rules and their actions are, within reasonable limits, usually predictable. On occasion they can depart radically from these rules and leave us wondering. The behaviour of two squirrels in October 1958 belongs to this order.

When first seen they were about to cross a road, the one being carried by the other, but in a very unusual manner. The two were head to tail, the one being carried clinging to the underside of the other, its tail held horizontally out in front [as illustrated in the drawing]. The squirrels crossed the road at a walking pace and disappeared through a hedge, but immediately afterwards were seen playing together in a tree just behind the hedge.

Grey squirrels are usually fairly solitary animals. They will come together in the pre-mating period, beginning in January, and will associate in groups for short periods during the day. Pairs will come together later for breeding. Otherwise the only associations are between parent and young. It is tempting to believe that the one observed was a squirrel bridegroom carrying his bride over the squirrel equivalent of a threshold, but it was the wrong time of the year for this.

69

Grey squirrel puzzle

There seem only two other explanations. Either the one being carried was a youngster that had stayed unduly long with the mother, and the carrying was an exaggerated expression of the maternal impulse to retrieve an infant. Or else the two squirrels had invented a new game.

PLAYFUL HOBBIES

In any claims for skill and grace in the air the hobby must be seriously considered. A summer resident, wintering in Africa, it is not a common bird with us although it regularly breeds south of the Thames. It becomes less common farther north; in Scotland it is no more than a rare vagrant.

About the size of a kestrel but with the flight more of a peregrine, seen at close range it shows a dark slate back and white underparts with black streaks. In the air its long scythe-like wings are an obvious feature. Perched, its wings extend slightly beyond the end of the short tail.

The fact that the hobby sometimes takes swifts on the wing is evidence of its speed and dexterity in the air. Hunting well into twilight it sometimes takes bats. But while hobbies may take a variety of birds, in addition to swifts, their food is mainly insects, such as dragonflies, moths and beetles—somewhat insignificant game for so skilful a flier. They are taken in the claws and eaten on the wing.

A strident repeated *kew-kew-kew* often calls attention to the hobby, perhaps in a group of pines. Then we may see a display of aerobatics, solo or in pairs. Stooping, tumbling and looping-the-loop may follow in rapid succession, and one of the birds may turn on its back and glide upside-down, apparently all in fun.

70

Hobby flying upside-down

MUSKY SIGNATURE

In more than one instance a thriving community of badgers has existed for years while people living in a village a stone's-throw away have been unaware of their presence. In the scientific literature, too, there are indications that their secrets are well kept. For example, it was not known for certain until less than 20 years ago at which time of the year badgers mate; then Ernest Neal, by close observation of their habits, was able to set the breeding season as July to August.

The pre-mating behaviour may include a great deal of play, even a stereotyped form of leap-frogging. It is also accompanied by the emission of musk, from glands beneath the tail. The other functions of these anal scent-glands are not fully understood even now. A badger will deposit its scent at various points in its territory, on stones and roots, even on bare ground, probably as a means of marking out the territory. It will do so, as well, when it wanders farther afield, and this it is believed is a means whereby it can find its way home.

Neal has suggested that a sow badger leaves her scent around in this way when she is ready to mate, as a signal to the male; also that she will do so when she has left small cubs in the set, as a warning to other badgers to keep away. If these suggestions are correct, then the language of a badger's scent must contain subtle nuances. Perhaps another is seen in the way a tame badger will treat a stranger. Having sniffed you it will turn around and apparently squat for a moment on your shoe, then walk away leaving its scent on you. Presumably this action has the same message as a handshake, a sign that you are accepted.

ACTING DAFT

People have sometimes reported seeing a fox prancing, bucking, somersaulting and rolling on the ground as if it had taken leave of its senses. Often such reports tell of rabbits and small birds gathering round, drawn out of curiosity, to watch the performance—and then the fox pounces. This behaviour has received the name 'charming'.

There have been similar reports for stoats, but for both animals the charming tactics are rare and little understood, so there has been much speculation as to their meaning. At one time it was supposed that the fox was being crafty, deliberately luring prey within reach. A more likely suggestion was that this was all a matter of sheer high spirits but that possibly when a fox found it paid dividends he was capable of using it deliberately on a future occasion to obtain a meal.

I once watched a fox indulging in such tactics in the snow. Against the unbroken white and with the light reflected brightly from the snow every detail became clearer. It was possible to see that the fox was playing with something. It proved later to be a piece of rotting squirrel carcase. The fox would throw it up and leap after it, or roll on it or cavort and buck around it as it lay on the ground.

It may be that charming is always stimulated by something similar but that sometimes the plaything is too small to be readily seen at a distance or against a broken background.

CHAPTER 7

HARASSMENT

BIRDS AND CATS

The death of a young bird at the paws of a cat is a distressing thing to witness. It is not so much, perhaps, the effect on the young bird which upsets us as the agitated behaviour of the parents. It would be folly to disparage the excellent sentiment that stimulates sympathy for the bereft parents, but it might help to consider such events dispassionately.

To begin with, it is as well to remember that if we keep cats such things are inevitable. They are also inevitable if we do not! For of every brood, the majority of young birds are doomed to early death from one cause or another. It is the same in all species but man. Indeed, if animal populations increased at the same rate as the human race, life for any would soon be impossible. The world would be overcrowded.

Now, as to the distressing cries of the parent birds, while we should not withhold our sympathy, it is possible to temper our own anxieties with certain matter-of-fact considerations. Are the agitation and the heart-rending cries of the adult birds signs of grief or merely a response to danger? Note how both are taken up by all birds in the neighbourhood, and die down in a short while once the danger is passed. Note the nonchalance with which the pipit accepts the loss of its own offspring at the hands of the cuckoo fledgling.

Animal cries do not always denote the same emotions as similar cries from human beings.

The setting was a square brick pillar in the full glare of the hot sun. A score or so of flies rested in a scattered group on its surface when a wasp appeared, flew at the largest fly and tried to straddle it.

Had the fly not moved away in the last split second, the wasp would have held it imprisoned between its legs, bitten off its wings and legs and sailed away with the dismembered carcase—in the

Wasp carrying away dismembered fly

way wasps do. However, the first victim had escaped, so the wasp took off, described a short circle in the air and came in again to attack.

This time half a dozen flies flew at it, buzzed it in a tight bunch and returned to their resting places. The wasp retired, circled and came in again to attack. Again it was buzzed, this time by nearly the whole score of flies. It retreated again, again was buzzed and again retired.

On the sixth attempt it gave up and flew away. The whole incident took little more time than is needed to read about it. A trivial incident, concerned with trivial insects, yet not so trivial in its import. One was reminded inevitably of the tactics of attack and defence in air warfare. Or, better still, of birds mobbing a hawk or owl, when several species will combine to harry the attacker. It was just one more example of the similarity in behaviour of widely different animals, poles apart in structure and mental equipment.

GNATS WATCH THE GIRLS GO BY

You seldom read much of the behaviour of male gnats. So far as we know, they feed on flowers, possibly nectar. And no scientist is needed to tell us what females feed on. They take blood, ours if we are about.

The way the female follows her dastardly course is to home on her victim in roughly three stages. When an excess of carbon dioxide, as from the human breath, reaches her, she takes off and flies up-wind. As she gets near her victim the slight increase in the temperature of the air, and the very slight increase in humidity left behind by a person, direct her more certainly towards her victim. Then finally her eyes come into play, she sees her victim and lands.

In this third stage there is an even greater concentration of carbon dioxide, and this must be a strong guide to her, because when we use chemicals to keep gnats away what we are doing is confusing the female by masking the carbon dioxide; she is put off course and swerves away.

Sometimes as we walk along we have a group of male gnats flying around our heads; some people call them 'following gnats'. What interest have they in following us? Can it be that this is a convenient way of finding a female? Do the gnats follow us on what passes in an insect for an assumption that sooner or later a female is likely to be attracted for her feast of blood, and one of them will then have a chance to mate with her?

SWALLOWS BAFFLE A HAWK

An Oxfordshire reader described to me how, at 4.30 pm on a rather hot day, in brilliant sunshine against a cloudless sky, some 200 swallows and martins were exercising, presumably before departure on migration.

Apart from their twittering, the only sound came from a noisy group of starlings. All of a sudden the starlings were silent, which made her look up, to see them rushing away to one side of the garden. Then she saw a sparrowhawk coming in from the opposite direction.

Within a split-second, the swallows and martins had formed a solid group and were advancing on the hawk. They kept on alternately dividing into two groups, one on either side of the sparrowhawk, their white undersides flashing in the sun, and forming up into a solid black oval ball behind the hawk.

The incident lasted only a matter of seconds, during which the hawk beat what seemed a leisurely retreat and was finally lost to sight surrounded by the cloud of pursuers. After their successful campaign, the swallows returned in one great swoop to continue their interrupted assembling flights.

The starlings and the swallows and martins were showing their inborn behaviour pattern in the presence of a hawk. Any beast or bird of prey relies on picking up stragglers or weaklings. The incident illustrates how, on migration especially, there is an obvious advantage in travelling in flocks, giving strength of numbers against possible attack all along the route.

MOB AGGRESSORS OF THE AIR

A flock of more than a score of crows wheeled and attacked, and then attacked again. Their quarry was in silhouette, but the broad wings with their slow beat, as well as the long trailing legs, identified it as a heron. Reporting this, a reader asks: would crows be mobbing a heron? Do they not reserve this behaviour for hawks and other birds of prey?

This is, of course, more usual, but it is not uncommon for them to mob a heron. The reason for attacking a bird normally harmless to them can only be guessed. Perhaps the large size of the heron alone is enough to arouse animosity, especially where the crows are in sufficient strength to build up a group courage. It is even possible that they do, in fact, mistake it for a bird of prey.

A third suggestion seems better to fit what we know of the temperament of crows. A number of birds react to fright, such as the shock of sudden attack, by regurgitating food. Other species have learned to take advantage of this to obtain a free meal. No doubt crows quickly learn that to mob a heron may cause it to disgorge a freshly swallowed fish or frog.

This form of parasitism is especially pronounced in the more belligerent sea-birds. Frigate birds have brought it to such a fine art that it is their main means of subsistence.

STARLINGS SCARE THE CAT

A correspondent described how, at 7 o'clock one morning, his attention was caught by a half-dozen starlings flying around one of his chimneys. Then he saw his cat, sitting amongst the chimney pots looking bewildered.

More and more starlings arrived and one by one they would peel off and dive down to within an inch or so of the cat. After a few minutes of this, the cat became so harassed, being unable to do anything to retaliate, that he jumped off the chimney stack and made his way down the roof. All the time he was descending the slope the birds followed, wheeling and diving. Only when the cat had leapt off the roof to a window sill and then to the wall below was the attack broken off. My correspondent added: 'I have never observed such concerted effort on the part of birds against an enemy, or potential enemy.'

Collecting observations on starlings and their defensive tactics has been a minor hobby with me for some time; but this is one of the gems. There is little doubt that, individually and collectively, starlings are outstandingly alert birds. They may bicker among themselves, but in a flash they can turn a common front to an enemy, or even a potential enemy—recognising, as in this case, a malicious intent. It is not surprising that starlings have multiplied exceedingly.

MOBBING THE OWL

Some years ago we took over a pair of tawny owls that had fallen from the nest. Both were slightly deformed and probably

78

would not have survived as free birds. The first morning after they were installed in a large aviary under a spreading oak the small birds gathered to mob them. The foliage of the oak seemed itself to be alive as the birds moved agitatedly in it; and the din was terrific. With each succeeding morning the din grew less and in a few weeks the mobbing died out. Occasionally a single bird, at most a few, would come over to scold the owls.

Regularly each night a wild owl from the woods opposite the house flew over to the oak to call to the owls in the aviary. For most of the year this happened after the small birds had retired. With the lengthening days, the wild owl came out while they were still about, and as often as not they would mob it. So here is the picture; the small birds seemed to have learned that captive owls are harmless, but they still mobbed the free owl.

It has been suggested, as the result of careful investigation, that the impulse to mob an owl is inborn and that the shape of the owl calls it forth. I am inclined to believe it is learned by each generation from its forebears.

THRUSTFUL THRUSHES

It seemed all wrong for a mistle thrush to be taking on a carrion crow single-handed: the thrush, one of the soft-billed birds, weighing 4 ounces and measuring 10 inches across the spread wings, against the crow at least four times its weight and measuring 2 feet across the wings, with a murderous beak in addition. Yet the thrush chivvied the crow relentlessly across the field and into a tall lime.

Seconds later back they came. This time, the crow, with an egg in its beak, dropped to the ground to devour it. The thrush still fought doggedly, flying back and forth over the crow, diving steeply on it each time, turning, banking and plunging. For all the concern the crow showed, however, the thrush might not have been there.

Nevertheless, tenacity in defence of the nest does not always go unrewarded, and it sometimes leads to strange encounters. A few days before the thrush, with its mate, had beaten off a jackdaw;

Mistle thrush attacking a carrion crow

and also a grey squirrel which they attacked so vigorously it was actually knocked from the tree. In view of the way a squirrel can hang on this was quite a feat.

Success boosts morale even in soft-billed birds. The mistle thrushes' finest hour came when, having vanquished the squirrel, three muntjacs came and stood under the tree. Nothing could menace a 20-foot high nest less than three deer but the thrushes, their monkey up, waded determinedly into the deer as well.

FALSE ALARMISTS

For a tawny owl to show itself by day is to invite harassment by a mob of small birds. Conspicuous among these are blackbirds and jays, if only because they are the most vociferous. In fact, it is more often than not the alarm notes of one or other of these, or of a magpie, that bring the small birds to the scene.

When, in June, these owls have their fledglings, they tend to move about more during daylight hours. A blackbird is usually the first to see them and give its rattling alarm cry. In a matter of seconds the cry is taken up by chaffinches, robins, thrushes and the rest of the songbirds, who fly in from the neighbourhood to crowd round the owl in a noisy chorus. If a jay or magpie should first raise the alarm, their raucous calls seem to have an even greater effect in bringing the mob together.

Last week the pattern of mobbing around our local owl family was unusual. Suddenly there built up a strident chorus of the harsh alarm notes of jays and the alarm rattles of blackbirds. This is most unusual. As a rule only one jay takes part and a few blackbirds, their cries welling up from among the cries of the smaller birds.

The truth, however, proved to be even stranger. There was the tawny owl at the centre, surrounded by a mob of starlings only. Instead of using their own alarm calls these famous mimics were rattling like blackbirds and squawking like jays.

CHAPTER 8

WHO'S BOSS

GUARD ON THE LARDER

More than a century ago Mayne Reid noted in his diary that the mistle thrush is a hardy bird that will even outlive winters which kill the fieldfares and redwings. He found this remarkable since these two species belong to more northern latitudes and should be expected to be able to endure cold in its extreme degree. He would surely have enjoyed the antics of a mistle thrush in a small suburban garden recently. He might even have seen a possible clue to his mystery.

The owner of the garden had put a small heap of brown bread cubes in the centre of the lawn. The resident mistle thrush took over. It sat on the fence and if any other bird so much as flew down to the bread it swooped and drove it off. Starlings had even less chance than the others, despite their usual aggressiveness.

All day long the thrush flew from fence to fence, driving away other birds and when evening came the food was still there except for a small quantity eaten by the thrush itself. Indeed, the surprise was to see how little it took compared with a starling.

The next day the owner of the garden put five separate heaps of food around the perimeter of the lawn, to see how the thrush would cope with the larger area to guard. By nightfall the five heaps were still there. A few sparrows had been allowed to take a few crumbs but any starling trying to do the same was seen off with a violent flurry of wings and loud cries. And that is how it still is at the time of writing.

A friend told me of two blackbirds fighting so much in her garden that she was almost tempted to go out and stop them. She thought perhaps food was scarce because of the snow and they were competing for it. I was able to assure her that if she had separated the combatants they would have fought it out elsewhere, or as soon as her back was turned.

With most birds fighting is mainly between males over territory, and on the whole the combats are more show than anything, a matter of bluff rather than inflicting damage.

Blackbirds seem to be more quarrelsome than most, the males fighting over territories and the hens fighting over the males. At least that is what we are told, but I have been watching blackbirds for some years and the impression I get is that they scrap for the sake of it. There was one, for example, that I had under observation, which for several days was beating the daylights out of a cedar cone lying on the ground. Every time another cock blackbird came near, he turned from his punch-ball and attacked the living adversary instead.

Certainly cock blackbirds fight more during February and March, when nesting is beginning and territories are being marked out. But my observations suggest that they are ready for a scrap at all times. The figures speak for themselves. There are many more records of deaths from blackbirds fighting among themselves than for any other of our native birds.

Cock blackbirds scrapping

83

At first there were two on the path, near to each other and not quarrelling, a fair sign that spring is not far ahead. Then a third flew down. When three robins come together there is bound to be trouble. Within seconds one of the original pair had sidled up to the newcomer, and the skirmishing began. To the casual eye it might have appeared to be play. The two would suddenly fly at each other, twist and turn in the air in a tight display of aerobatics, then separate.

Perhaps the slickest turn occurred after one of these bouts, when the two were separated by about six feet. One dropped to the ground and stood with his head thrust forward. The other took off, planed down and, closing his wings, passed immediately under the other one's chin, neatly 'buzzing' him.

It was not play; that much had been clear from the moment the third robin appeared. His opponent had sidled up to him, raised his head, and thrust the beak vertically into the air, as if straining to look at something high up in the sky. Thus, with the body held rigid, he swayed in slow motion from side to side. To the human eye the action was comic to the point of being ludicrous. To the bird it was the first challenge, the typical aggressive display of a cock robin on his own territory with his own mate, warning off an intruder by displaying his scarlet front to the full.

COURAGEOUS COCK BLACKBIRD

The jay may or may not have had designs on the nest. It seemed to be making persistently in its direction, but the blackbird attacked with even greater persistence, jostling the jay when it went into the tree or alighted on the ground, and swooping across it every time it took to the air. In the end the blackbird won.

It is often said that jays and magpies are destructive of small birds. This may be true; but although there are many jays and magpies in our own district, small song birds are still abundant there. A predator tends to weed out the sick and the weaklings, as well as the young of parents lacking the courage or ability to defend them. On the whole, the chances are in favour of courageous parents begetting courageous offspring. A balance is thus maintained as the work of the predator is kept in check by a natural increase in the courage of its prey.

Courage, especially in defence of the young, is proverbially associated with the female, but in this instance it was the cock. Normally shy and apparently timid, quick to raise the alarm when his brood was threatened, he showed a disregard for his own safety which eventually rewarded him.

It seems almost a rule that those preying on others have a streak akin to cowardice. One is continually seeing instances, even in the animal world, in which willingness to stand up for oneself has surprising results in a contest against heavy odds.

LIZARDS IN ANGER

A group of lizards taking the sun on a bank consisted of females and half-grown young from the previous year. The scene of tranquillity and immobility was shortly disturbed by a rustling under the dried leaves to one side.

Finally, two male lizards rushed out. The pursued turned to face his pursuer. Face to face they lowered their heads, arched their backs and raised themselves with all four legs fully extended, absurdly like two cats sparring for a fight.

Suddenly one lizard sprang at the other and appeared to seize him by the neck with his jaws. Violently they rolled over and over, a fury of scrambling feet and waving tails, until they broke away to face each other once more. They sprang at each other again, each seizing the other by the jaws. Finally one gave up the fight and scuttled for cover.

The breeding season for lizards was drawing near and it could be that the sunshine had brought these two early into a fighting mood. If they were fighting for a mate she kept herself well hidden; the other female lizards seemed totally indifferent.

ADDERS' DANCE

Reports of a 'plague' of adders near Swanage in 1957 drew comments from several readers about the numbers seen in other localities. One tells how last year he was compelled to make a number of journeys to the vet to have his two Siamese cats treated for snake bite. The next year a third cat, an ordinary black-and-white stray that had adopted his family some months previously, turned the tables somewhat. It killed an adder, biting it through the spine about three inches behind the head, without itself receiving injury.

The probable explanation of these apparent 'plagues' is that in February and March, when they leave their winter quarters, the adders stay in the neighbourhood, a dozen or more coiled up together. In April and May they breed, and there is much excitement among them.

One expression of this is the curious dance in which the territorial rivalry of the males often culminates. Two males face one another, raise the fore-parts of their bodies, and sway from side-to-side. Then as they come together they push and thrust, parry and duck, entwining their bodies in an effort to throw each other to the ground.

The contest may be broken off and renewed several times, until finally one adder departs, ceding the victory. There is no biting, only intense excitement. The female lies coiled up near by.

BATTLING YAFFLES

A correspondent from Christchurch, Hants, sent me the following story. Thirty feet from his bedroom window is a pine. Woodpeckers have for years worked at a hole in the trunk. This year a pair of yaffles (green woodpeckers) worked at it until one of them could get in, but had to back out.

One day one of them arrived at 5 am and worked continuously until 8 am. Then the second appeared and they mated on a nearby branch. Both then flew off and a great spotted woodpecker came to dig. When he had gone two starlings came. One went inside.

Presently a yaffle came back and began furiously pecking the bird in the hole, dragged it out by the scruff, carried it a few yards and dropped it. The starling flew away followed by the yaffle in a chase lasting 15 minutes. The yaffle then entered the hole leaving

Green woodpecker

the tips of his tail feathers protruding. Three starlings then arrived and started to tweak the tail until the owner began to back out, when they beat a retreat.

The performance was repeated each morning. Once an intruding starling was killed by the yaffle's sharp beak.

The yaffles took turns in guarding the hole, but sometimes both left it and the starlings took possession, taking in nesting materials which the yaffles later threw out. On subsequent occasions when a woodpecker was perched near the hole one or other of the starlings would fly near it. As the yaffle chased it another starling would slip quickly into the hole, only to be ejected on the yaffle's return.

Eventually the starlings won.

YOUNG WATER VOLE CHASTISED

The water vole was swimming upstream with her quarter-grown youngster following in her wake. Suddenly there was a flurry of water, squeaks from the youngster, and in no time, the skirmish over, the two went swimming on.

Now, however, the youngster, instead of being immediately behind its mother, was behind and slightly to one side, and showing every sign of being ready to dodge out of reach.

It is a moot point how far birds and beasts actively contribute to the training of their young. In birds, especially, there is good reason to believe that most of the training is by example. In some birds the first 48 hours after hatching are all-important in their education. In domesticated animals, particularly in dogs and cats, we can, however, see signs of a more purposive training, although undoubtedly, as with human beings, training by example is still the most important. If corporal punishment is used, and it seems to be on occasion, it is used infrequently and with discretion.

It is almost a scientific sin nowadays to make close comparisons between human and animal behaviour. And the skirmishing between the two voles may have had an entirely different significance. Nevertheless, there was something familiar in this scene, as the young vole followed its mother upstream, just as a child, having been recently chastised, follows its parent just out of arm's length and ready to dodge.

PACIFIST SHREWS

Shrews have always been looked upon as quarrelsome. If we may judge from a recent scientific investigation, this is a libel.

A shrew feeds on insects. Because it requires a great deal each day, and its food is sparsely scattered, it cannot afford trespassers. So if one shrew meets another it squeaks at it to warn it off. The situation then begins to resemble that in which two cars have met head-on in a narrow road. If one gives way all is well. If not, the motorists sit tight and hoot at each other.

If one of the shrews refuses to give way, both squeak at each other. After a while, as a rule, one grows weary of this and departs. If not, one of them rears up on his hind legs, still squeaking. Should the intruder still refuse to take the hint, they do not fight. They throw themselves on their backs and lie there, squeaking furiously.

This is very like the habit, said to have been indulged by some humans, of biting the carpet in a rage. That takes place presumably when the person is on his own. With shrews, it is an instrument for settling quarrels without resort to bloodshed.

At least they set an example to their betters. But many birds also settle their differences by a singing contest.

CHAPTER 9

COLOURS INTROVERT AND EXTROVERT

COLOURS USED AS WARNINGS

The struggle to live in the wild is tough and unceasing, and all animals, large and small, find a measure of security in being able to efface themselves. Either they are blessed with a colouring which makes them inconspicuous in their natural haunts, or they habitually keep out of sight, taking advantage of available cover. There is, however, a small percentage that is not only strikingly coloured but makes no attempt to hide.

Indeed, they seem to flaunt their colours—and well they may, for they usually possess poison glands or stings, or are generally unpalatable. It has been assumed for a long time now that these bright colours are warning colours, a sort of 'touch me if you dare'.

Of recent years tests have been made to see whether this is correct or not. Our common ladybirds furnish a case in point. They never hide and their bright red-and-black coats cause them to stand out conspicuously. Somebody once went to the trouble of counting the number of insects fed by a pair of starlings to their young brood. The number totalled nearly 17,000; they included all kinds of insects, but only two were ladybirds, although there were plenty about.

Other counts have given similar results, and there is little doubt that the unpalatability of the ladybird is linked with its conspicuous colouring to give it an almost complete immunity from attack.

Striking, too, is the fact that the only poisonous lizards have

89

Ladybird and larva

conspicuous colouring to give them almost complete immunity from attack.

BLACK AND YELLOW FOR DANGER

The near approach of a wasp can strike something approaching terror into some people; and none of us is without a feeling of apprehension. Yet a wasp is more concerned with minding its own business than with committing acts of aggression.

It seems there is something about its vivid black and yellow colour that suggests danger. So far as human beings are concerned, this is to the wasp's disadvantage, for we lose no opportunity of swatting it. With insect-eating animals, however, it spells an almost complete immunity. This immunity is conferred, moreover, on many other insects bearing the same brilliant colours. Insects such as hover-flies and wood-wasps are left severely alone by insectivorous birds, although they are inoffensive in the extreme. It is the yellow and black that does it, for apart from this there is little in their habits, manner of flight, and so on, to justify confusion between them and the sting-bearing wasp.

There is, however, one insect, the wasp-beetle, that takes its mimicry further than a mere copying of the colour. A true beetle, it has a most unbeetle-like wasp-waist to its black and yellow pat-

Wasp beetle and wasp

terned body. It also has the same alert, jerky movements as a wasp, and the same quivering antennae. To complete the deception, it haunts wooden posts and fences where wasps are apt to go to gnaw off fragments of weathered surface for use in making paper to build their nests.

MURDER ON THE FLOWER BEDS

Crab-spiders spin no web, they crouch motionless in the centres of flowers. But no insect could be expected to visit a blossom, no matter how strong the call of the nectar, with a spider sitting on it. That is where this particular spider, with its two pairs of long front-legs, has a pull.

The crab-spider can, within limits, change its colour to suit its background. On a pure white flower it is white. Place it on a yellow flower and it will change to yellow; on a greenish-white flower and it becomes greenish-white. The change is slow, it is true, taking up to 48 hours; but then the spider does not often change its station. Endowed with endless patience and the ability to fast, it can afford to wait. When it kills there is no fuss, no disturbance; its victim is transfixed in a natural position. There is no change in position to betray the spider's camouflage.

An experimenter once set out a number of dandelion heads, some with a small dark pebble placed at the centre, the rest without. He watched the bees and flies visiting the flowers. With rare exceptions they avoided those with the small dark pebbles. A strange instinct, surely. We say it has a survival value, but this merely covers up our complete ignorance of how it first came into being.

But we have here some measure of the efficiency of those large insect eyes, with their many facets. Though they can see the obvious, subtleties of form and shade escape them. In fact even the keen human eye seldom detects a crab-spider until it moves.

CATERPILLARS AT HIDE-AND-SEEK

Looper caterpillars are adepts at the game of hide-and-seek. At least they are at the hiding part of it. We have been keeping watch on one for some days in a small apple tree. This was far from easy with a caterpillar that looks exactly like a twig, especially when it can move surprisingly quickly from one part of the tree to another. The trail of leaves with lumps chewed out of them was more obvious than the one that did the chewing.

There is always the possibility that, seeing things through human eyes, we deceive ourselves. Birds are one of the caterpillar's enemies and their sight is much keener than ours, about ten times as acute. It may be that some of these camouflage tricks in animals do indeed fail to deceive a keener sight.

At least it was worth while putting it to the test, so the looper was closely scrutinised under a strong lens that magnified it not 10 but 20 times. It passed the test. The body was a slaty-grey with a tinge of reddish-purple and marked with fine, irregular streaks, very like the bark of a real small dead twig. It was irregular in outline too, almost gnarled.

It bore the same minute yellow spots as we find on a twig, and where the caterpillar's hind end grasped the twig was hardly different, even in detail, from the point where a real twig joins a stem. Above all, the caterpillar's face and front legs resembled to an amazing degree the distorted end of a dead twig.

PUZZLE OF A DABCHICK

The half-grown dabchick did not dive, as might have been expected, but swam rapidly upstream to disappear into the bank. A detour brought me to a point on the bank opposite where the bird might be. It was there, sure enough, with its back to the rotten roots of a tree stump in the overhanging bank.

It was sitting in the water facing me, and, against its background, virtually invisible. I watched it closely for half an hour, from a distance of eight feet, until the cramp from squatting drove me to move. Throughout this time the dabchick made not the slightest movement that I could detect, even when I moved suddenly or violently. Only when I had gone (to hide behind a bush) did it move, to swim rapidly to where, a furlong on, the stream joined the river.

There seems little doubt that the dabchick deliberately chose a spot where its plumage blended perfectly with the background. And as if fully aware of this, and confident of its protective value, it remained dead still. Was this a piece of automatic behaviour, devoid of thought? Perhaps it was; we do not know enough about these things to answer positively.

The dabchick normally lived on the river, and was off its beat in this small stream. The river is several feet deep, and the dabchick automatically dives when alarmed. The stream is a few inches deep, and the dabchick made no attempt to dive. Instinct or thought?

LEAVES COME TO LIFE

One of the first evident signs of spring in England and Wales is the sight of the brimstone butterfly on the wing. It is one of the few of our native species that passes the winter as a perfect insect. Whereas the others, such as the peacock butterfly, overwinter in hollow trees and other crannies, in outhouses or even within dwelling houses, the brimstone spends its time in an evergreen bush, among the foliage. There it is inconspicuous because the undersurfaces of its wings have an almost perfect resemblance, in shape and colour, to a yellowish leaf.

The brimstone inhabits thickets and hedgerows and can be seen flying strongly along woodland rides, wherever the buckthorn grows. It is on this the female lays her eggs, which are pale green at first, changing later to yellow, and fixed usually on the underside of the leaf, on the mid-rib.

The caterpillars, best seen in June and July, are green, almost exactly the colour of the buckthorn leaves. When ready to pupate, each caterpillar fixes itself by its rear end to the underside of a leaf. Then it spins a loop of silk around its body, the ends fastened to the leaf, before turning into a greenish-blue chrysalis, again looking very like a leaf.

The male brimstone is sulphur yellow, the female greenish yellow. In both, the tips of the fore-wings are sharply pointed. There is a central orange spot on each wing and reddish dots on the outer margin. The antennae in both sexes are reddish.

VANISHING ROE DEER

In Neolithic Britain roe deer were everywhere. Today they are truly wild only in Scotland and parts of northern England. Farther south they are also numerous, but there they are the descendants of deer deliberately introduced that went wild.

The roe, an elusive animal of the woodlands, has the trick of lying up in dense cover and coming out mainly at night to feed. So you may walk about in a small wood for hours by day, where up to 30 roe are known to be, and yet have only the occasional fleeting glimpse of one of them.

The trick of concealment is inculcated from birth. The fawns, often twins, are born in May or early June. They are able, like other hoofed young, to walk within a few hours of birth and to run with the parents within days. For a while, however, their hope of survival rests in lying hidden.

As if aware of this the doe forces her fawn to squat in cover, soon after it is born, by pressing on its back with her foreleg. The fawn quickly learns to lie crouched, to stretch its neck along the ground, to lay its ear flat, and to hold this position without moving so long as danger is near. Meanwhile, the doe moves off, not because she is deserting her youngster, but to draw the enemy farther afield. It is a piece of teamwork, persisting over the centuries, even after the natural enemies have been largely eliminated.

BULLFINCH IN HIDING

The snowberry is an alien, naturalised long enough to pass for a native, familiar yet unnoticed. An ordinary-looking shrub, even its pink flowers are too small to catch our attention. When the leaves have fallen, however, its tangle of slender stems show up russet red, and its berries like small snowballs stand out strongly. They catch the eye, and they catch the eye of the bullfinch.

We are often assured that the colours of a bird's plumage are a camouflage, and this is strikingly true in some instances. The nightjar harmonises so completely with a background of moss, lichen and bracken as to be almost impossible to detect on the ground. But we can make too much of this idea of camouflage. There may be a dozen birds in the same piece of shrubbery, all coloured differently yet all difficult to see, even those with the bright colours. The best camouflage is to keep still, or to fly away out of sight.

A bullfinch readily flies away, very shy, as if aware that its colours are no help. The black cap and, more especially, the white rump, of both sexes, and also the red front of the cock bullfinch, do not help concealment. When foraging on the snowberry, however, the general colours of the plumage make a close match with the reddish bare twigs of the shrub, and the white rump simulates almost to perfection the shape and colour of the berries. A perfect camouflage, one might say, as if they had grown up together, yet

95

the bullfinch has been here for thousands of years and the snow-berry was brought across the Atlantic only a century ago.

Nightjars are mainly heard and not seen. They fly at twilight, not the best time for seeing, and by day they rest on the ground, their grey-brown plumage, mottled and streaked, making them difficult to see against almost any background.

Their call is a churring note, unmistakable once heard, which may go on without a break for five minutes or more, rising and falling slightly at times, and audible in the still of night a mile away.

This year the nightjars on the heath away to the south from here seem to have been unusually silent. I was beginning to think the birds might have forsaken this area when news came of a nightjar having made an emergency landing in the playground of the village school. It proved to be a juvenile, unfortunately moribund, but at least it suggested that the nightjars still breed here.

This youngster, like all of its kind, demonstrated as it rested on the ground, the remarkable camouflage effect of the plumage. As

Nightjar

96

if confident in its efficiency a nightjar remains immobile until you almost have your hands on it. Then it takes off with incredible suddenness. The bird even closes its large eyes to slits so that they do not act as give-aways.

Moreover, all abrupt movements are avoided or masked. While at rest a nightjar may need to scratch or preen. It starts to sway the head and fore-part of the body, so that it looks more like dry vegetation swaying in the wind. And then, having scratched, it sways gently again before coming to rest in its typical and utter immobility.

CHAPTER 10

HOARDS AND CACHES

MOLES AND DROUGHT

What happens to moles during a drought? Like so many small insectivores, a mole is always on the move, working feverishly day and night. And feed it must, and keep on feeding, to maintain such a high pitch of ceaseless activity. Four hours without food will kill a mole. Although they are classified as insectivores, the main diet of moles is worms, and in dry weather worms go deep, to 6 feet or more below the surface. Moles' powers of tunnelling and digging are amazing, and everything about them is subordinated to this task.

They can enter the ground so quickly that they may be said almost to dive in. They can skim through the ground just below the surface at a rate comparable with human walking. But they cannot go so deep as 6 feet; or if they can they do not make a practice of it.

Therefore, in times of plenty moles, like so many other animals, must store food against an emergency. What then? Kill worms when they are abundant, store them in their runs and underground chambers, and have them go putrid later and foul the runs? The answer is simpler than that. Up to a certain point worms can re-grow lost parts. In wet weather, when there are more worms in the surface layers than even a mole with his tremendous appetite can manage, he will bite the heads off those he cannot eat, so that they cannot burrow, and store them in underground chambers.

If abundance continues, his stored live food re-grow their lost heads and tunnel away, living to be eaten another day. A store like this will last two weeks, perhaps three.

It is highly probable that 99 out of 100 people, asked to name an animal that stores food, would name the squirrel. Yet there are few, if any, animals that do not store food, either as fat within the body or in actual larders or granaries. The annals of natural history teem with examples, even among our native animals, many of them far more remarkable than anything we know in a squirrel's behaviour.

Rooks and crows store walnuts, water voles store grass, moles store earthworms; and so on. Perhaps even more extraordinary is the neatness often shown in the process. Stoats are especially noted for arranging their caches in neat piles. They may consist of rats, mice, frogs, birds and other such items arranged in rows, usually with the heads pointing in one direction.

How far this tidiness contributes to the welfare of the animal, indeed how far the food store itself is used to advantage, are points upon which there is little positive information. It may be no more than a twist of behaviour having little practical value, for we are not sure that all such stores of food are eventually eaten.

Can we suppose, for example, that the stoat which made a neat cache of golf balls really mistook them for food, or was it, like philately in human beings, the result of a pure urge to collect?

A GARDEN THIEF UNCOVERED

It is an old story, of the man who hid a half-bushel bag of hazel nuts in a shed. When he came for them a few weeks later there were only a few at the bottom of the bag. Fortunately for those who came under suspicion, all the nuts were found in various jars and tins lying about the shed, crammed to capacity. The culprit, too, was found—a wood-mouse.

Once, digging in the garden, I unearthed a pint of holly berries. A few minutes later, within a few feet of the hoard, I disturbed something in a pile of vegetable litter. It made off into cover in a few 2-foot-long leaps, like a frog, allowing me no more than a quick glimpse of a reddish-brown back, long hind feet, prominent pricked-up ears and a trailing tail. That also was a wood-mouse. Now I know why my garden peas sometimes fail.

The wood-mouse or long-tailed field-mouse is a picker-up of such unconsidered trifles as nuts and berries. It will also take peas and beans, bulbs and corms, wheat-grains and the like even when planted out. On the other hand there are several points in its favour. Charming and inoffensive, it lacks the strong smell of the

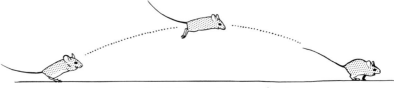

Longtailed fieldmouse kangaroo-leaping

house-mouse. Should it come into a house, which it does but rarely, it does not gnaw its way in, but will use holes and crevices in the brickwork.

HEDGEROW BUTCHER

The red-backed shrike reaches us in early May and leaves again before September. It is local in distribution and seldom breeds north of Cheshire and Yorkshire. Where it does settle, however, it is easy to see, especially the cock, with its distinctive red back and grey head. Other marked features are the black eye patch, with a white streak above it, the hooked beak, rosy buff underparts and tail with white margins.

Although little larger than a sparrow it has something of the dash and ferocity of a hawk. It takes up position on a favourite perch and from thence makes sorties to capture insects, lizards, mice or small birds. While perched it watches keenly for every movement, turning its head from side to side, looking down or up, or even turning the head to the rear and then looking up, so that its chin is towards the tail. Its prey when captured is held in the foot. Wasps or bees, or large beetles, are also held in the foot and hammered on the ground to beat the life out of them.

The favourite haunt is a hawthorn hedge, and over this the shrike will beat, like a hawk, poising with rapidly beating wings before gliding on to poise over another likely spot.

The thorns of the hedgerow are put to an unusual use. On them the shrike will impale its larger victims before dismembering them, collecting the carcases in a so-called larder. Hence its other name, butcher-bird.

WHAT MARKS THE SPOT?

Although squirrels hide nuts and acorns it seems likely that they often fail to return to their stores. A jay or a magpie, by contrast, will go straight to a hidden acorn, or at worst will search the ground nearby until it finds it. The question is, how do they know?

Several birds hoard food. Chief among them are crows, rooks, magpies and jays, but coal tits and marsh tits do so as well. Rooks and jays are not merely content to bury food near where they find it, but will often carry it far away, up to a mile. In rooks there is something almost suggestive of foresight since in autumn they will transport acorns and bury them on the winter feeding ground.

A jay about to cache an acorn

Squirrels returning to buried nuts are supposed to smell them out, but birds are believed to have no sense of smell worth talking about. Therefore it is assumed that they memorise the surroundings when they choose a cache for food. That seems the only reasonable assumption since they come back to it days, weeks or months later and go straight to the spot in spite of changes in the vegetation.

Rook with throat pouch filled

Anyone who has buried a small object without marking the spot precisely will recognise this as a truly remarkable feat of memory—if that is what it is, for the mystery does not end there. Birds that have been seen to bury food have also been seen to go straight to the cache again when there are several inches of snow on the ground.

FOR A RAINY DAY

Squirrels have the best reputation as hoarders of food, but some species of mice and rats are even better at it. Indeed, many times stores of nuts and acorns unearthed in gardens have been supposed to be the work of squirrels when, in reality, they were the hoards made by long-tailed fieldmice and yellow-necked mice. There are some birds, also, that run squirrels a close second, notably marsh and willow tits and coal tits, and in Scotland the crested tit.

A reader told me of a pair of coal tits that come regularly for peanuts. She had noticed they came so quickly (about every 15 to 20 seconds) that they could not have had time to eat them. By watching she found they were hiding them on low walls and in crevices; and she wondered if these were for future use.

Coal tits, about the size of a blue tit, have an olive-brown back and a black cap but with a white nape. The marsh tit is about the same size, a brownish bird with a black cap, the black extending down into the nape. Both of these, as well as the willow and the crested tits, store food, usually in crevices in bark. Tests with coal tits have shown that each bird will return deliberately to its own caches. Moreover, it will find the beech mast, pine seeds and other items pushed into the bark even when these have been masked so that they are no longer visible, showing that they recognise unerringly the places they chose for their titbits.

PRUDENT ROOKS

While watching rooks in a field a short while ago I saw several of them cache food. I remarked on this to somebody who came and stood near me. She told me how, when living in a house with a garden looking out on to arable land, she had seen it day after day.

She had made a habit of putting out bread, and soon the rooks matched this by coming each day to take it. She soon noticed that at the end of a feeding session the rooks would take what remained of the bread and bury it along the edge of a herbaceous border. Then it began to dawn on her that there was more to it than this. She noticed that when, the following day, the rooks flew in to take their rations for that day, the first thing they did was to dig up the food buried the previous day and eat this. Then they set to work on the fresh supply and, at the end, buried the remainder.

If correct, this is a singular example of prudence. Yet there is no reason to doubt it. Rooks have been noticed burying food, but nobody that I know of has made a habit of training them to come and feed almost on the doorstep.

It could be that it was a habit developed by this particular flock, a sort of local tradition, in response to the efforts of a benefactress. More usually, from what I have seen, rooks bury surplus food to prevent their pals from having it.

CHAPTER 11

SURVIVING THE WINTER

A FALLACY ABOUT WINTER

Romantic writers sometimes enlarge upon the sorry plight of our wild birds and beasts in winter. They draw attention, with more sentiment than knowledge, to the fact that while human beings can change from summer to winter clothes, animals cannot.

But this, surely, is a fallacy. There is a marked change, for example, in the red squirrel, for at the onset of winter we see it with a thicker body fur, a more bushy tail and longer ear-tufts. The change is not so obvious in the grey squirrel, but it, like all animals of the temperate regions, grows a thicker fur. In the colder, more northerly regions, of course, many animals change to a white winter coat. It is not known for certain whether this change results in greater retention of body heat. At all events, it is doubtful whether it has any other protective value. It is usually assumed that the new coat makes the beast invisible against the snow, but since almost all its enemies also change to white coats, any possible advantage seems to be cancelled out. As to birds, it is known that some, like the grouse, have additional feathers in the plumage in winter.

Whether this is general or not, all can puff out their feathers—which, by all accounts, has the same effect of retaining bodily heat. Even if birds do suffer slight discomfort in severe weather the chief danger for them is shortage of food.

BIRD THAT SINGS IN STORMS

To make progress in that wind one had to lean forward and push against its gusts. Overhead the gulls were almost stationary as they headed into it, and starlings flying vigorously were carried sideways like blown leaves.

Above the howl of the wind through the branches came the sweet notes. They might have been those of a blackbird, thrush, or robin, but there was no doubt where to look for the singer—in the topmost branches of the tall bare trees.

The wind bent the branches over so that the mistle thrush

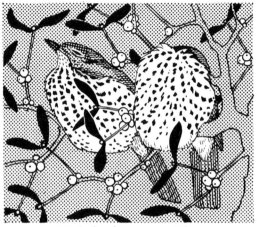

Mistle thrushes among mistletoe

swayed, perilously as it seemed, first tail up, then head up, as it clung to its perch and sang its snatches. The mistle thrush, deservedly named the stormcock, sings in the bleakest of weather, in rain, wind, sleet or snowstorm, and always in the topmost branches.

Its toughness extends in other directions, for it is said to eat the mistletoe berries, which other birds find repellent. It is the one that, traditionally, wipes its beak on the bark of a branch, thereby planting fresh seeds of mistletoe. It takes all manner of strange berries, and one result of the fashion of planting ornamental shrubs, especially those bearing berries or soft fruits, is a marked increase of mistle-thrushes in suburban areas or cultivated parklands.

They would, perhaps, be even more numerous but for the way

they succumb to cold weather. By a paradox, the stormcock that defies the elements dies in large numbers during a severe winter.

CASUALTIES IN COLD WEATHER

A mild winter does not necessarily mean more insects in the following summer. Whether over-wintering as eggs, larvae, pupae or adults they are capable of surviving remarkably low temperatures.

What is more likely to make an impression on their future numbers is the extent to which they are eaten during the non-breeding months. Given mild weather, the ground is more open for the foraging of insectivorous birds. In addition, small mammals such as various kinds of mice, as well as shrews, will be more active. Several of these do not hibernate in the strict sense, but very cold weather forces them underground.

Cold weather seems to be more damaging to the small mammals than to insects or even to small birds. The fieldmice, voles and shrews, all of which take insect food to a greater or lesser extent, suffer from what is known as cold-weather starvation.

With a body of such small bulk there is a proportionately greater surface for the loss of heat by radiation. In normal circumstances a high proportion of their food serves merely to make good this loss; all is well until the supply fails.

Low temperatures, with frozen ground making the search for food more difficult, catches them in the vicious circle of greater loss of body heat in searching for food combined with less food to be found. Continuing hard weather means a heavy mortality, the evidence for which is mainly underground.

SLUGS THAT DO NOT MIND THE COLD

Slugs are not necessarily inactive because the weather is cold. Some are merely out of sight; the field slug, white or pale yellow mottled with black or brown, that eats the potatoes underground and attacks the leaf crops at the surface, is one of them.

It breeds and grows throughout the year even when the temperature is near to freezing point. Pairing may take place at intervals of three to four days. The eggs hatch in three to four weeks and the young start to breed at three months, the normal life-span being 18 months. Someone once kept a pair of these slugs under observation for a year, during which they laid 800 eggs. Assuming all these had hatched successfully, reached maturity and started in turn to breed, this pair of slugs at the end of the year would have been the grandparents many times removed of several thousand million descendants.

Slugs have many enemies to keep the balance. Certain minute worms and other small organisms in the soil take up the offensive early and attack the eggs. From then on to maturity the growing slug is beset by enemies, from fly larvae to several kinds of beetles, from birds, moles, rats and voles to hedgehogs.

Slugs are classed as vermin, which is alone enough to give them a bad name. Yet they are among our best composters. Sometimes they make mistakes and compost the wrong thing; but for that, we should have nothing but good to say of them.

WINTER BUTTERFLY

Late one October, writes a reader, a butterfly flew round the electric lamp and settled on the wall in her flat. There it stayed until the end of December, when it was accidently knocked down and lost sight of. In January it reappeared fluttering across the floor. My correspondent tried to pick it up, but the insect's attempts to escape and her own fear of injuring it resulted in failure. It solved the problem by flying up and taking a sun-bath on the lamp. Then it flew round the room for an hour, settled on the piano and went to sleep.

Its description shows it to be a small tortoiseshell. The writer comments: 'I thought a butterfly lived only 24 hours. How did this one live without food or drink?' Some butterflies may be short-lived but many survive for long periods. The small tortoiseshell may become quiescent in some sheltered spot as early as late July and remain there until spring.

This is called over-wintering rather than hibernation. The small tortoiseshell, the rare large tortoiseshell, the peacock and the rare Camberwell beauty over-winter in hollow trees, outhouses or even dwelling houses.

BACHELOR GANGS

During spring and summer no bird is more tied to its territory than the chaffinch. All who have studied it comment on the way it keeps to a daily routine, visiting the same tree and the same branch, and doing so with almost clocklike regularity.

At the end of the summer the pattern changes completely. The chaffinch becomes a wanderer. Another big change takes place at the same time: from being intolerant of his fellows the cock chaffinch seeks their company, and so do the hens. This ganging up into flocks is not unusual. Many birds are more sociable outside the breeding season, particularly in hard weather.

But even in the more open weather of autumn, linnets, greenfinches, sparrows and buntings assemble in flocks, often several hundred strong, to forage across the fields. Sometimes the flocks are very mixed, and they may include not only various finches and buntings, as well as several kinds of tits, but also on occasion winter visitors like the brambling.

Perhaps the most remarkable thing about the chaffinch is that the sexes so often form separate flocks, the hens in one, the cocks in another. This is by no means the rule, but it is frequent.

The hen chaffinch is less brightly coloured and therefore less easily detected among a crowd of sparrows or other small birds in mixed flocks. As a result it was once thought the hens migrated south, leaving the cock chaffinches behind—which earned the cocks the name of bachelor birds.

LARKS WINTER ABROAD

On my winter walks across the fields, the skylarks have retreated before me, a half-a-dozen at a time fluttering up a foot or two before dropping back into the grass. When snow lay on the ground they were not there, which is not surprising, as their food is mostly insects or seeds.

They say that larks go westwards when hard weather comes, or to the margins of marshes where the ground is still soft. We who have not had the opportunity to see for ourselves must take their word for it, and for the statement that there is a more widespread general post. Bird-ringing and other forms of mass-observation show that skylarks breeding in this country are not those that flutter up at our feet in winter.

Solitary in the breeding season, skylarks congregate in groups for the rest of the year. A few remain through the winter but numbers emigrate southwards in autumn and return towards the end of February. In autumn also large numbers of immigrants from the Continent reach our north and east coasts some as winter visitors, others passing through to go south. The bulk of our winter skylarks are therefore visitors from abroad.

The number of larks on the move at such times must be large. When the reprehensible practice of trapping them was in fashion, a century ago, to catch a thousand in one night was quite usual.

A QUEEN'S CHOICE

Wasp colonies are breaking up and the workers dying, but more wasps will be back next year. The new queens are just beginning life by going into a state of suspended animation. Having mated they are now seeking a hide-out until next spring, a secure place where they can sleep the sleep of the dead until the time comes to found new colonies.

A great deal of insect behaviour can be reduced to simple mechanical terms. We can show that when the temperature falls they do this; when it rises they do that. We can, in fact, analyse their actions until they appear to be little better than animated machines. Yet this is not the whole truth.

In the south of England, when winter days are mild, the queen wasps can be seen inspecting the cracks between the tiles on the roofs, looking for a place to hide up. Cold is not driving them to it. They are anticipating a future need, impelled by an urge which we little understand even if we try to hide this ignorance by giving it a long name.

But there is more to it than this. The queen wasp searching the tiles tries first this crack, then that. She takes a long time settling where to go in for the winter, as choosy as a woman in a hat-shop or a man buying a tie. And like them she has the power of choice.

KILLER IN THE SNOW

Reports of sheep in deep snow having been savaged by foxes are an indication of the desperate straits wild animals find themselves in during hard weather. So are the footprints which show how, in other parts of the country, foxes have been going from one litter basket to another, scavenging what they can.

Those who can look at the situation more objectively than the

Prowling fox

sheep farmer can profit by the information revealed by snow. Although the common or red fox is so familiar, and ranges over the whole of Europe and much of Asia, as well as part of North Africa, our knowledge of it is relatively sparse.

It is sometimes said that a fox will travel long distances when out hunting. By studying the trails in the snow we get a better idea how far this can be. Over the North Downs there were clear footprints of a fox in the snow along the crown of the road. They were continuous for two miles, and this was only part of the homeward journey. From them, and from other signs, this vixen must have covered not much less than 10 miles in a night.

A common sight is the place where a fox has been pouncing on field mice. This curious and exaggerated action consists of rearing up on its hind legs and then plunging forefeet and muzzle into the snow. At another place the trail of a fox ceased abruptly and began again 15 feet farther on—a long-jump record?

Snow was thick, but still some green vegetables stuck out above it. The woodpigeons soon remedied that. And then they turned their attention to the holly and the hawthorn. Tree after tree was visited in turn and stripped of its berries.

There is something bizarre about a well-grown holly tree festooned with a score of woodpigeons. They perch at all angles, usually with wings half spread and fluttering, to maintain their uneasy balance on the yielding foliage. On the hawthorns, with their berries at the ends of slender twigs, the sight is more comic as one pigeon after another hangs upside down, like an enormous blue tit, to fill its crop.

Apart from the damage to the frozen green crops, large flocks of these pigeons must seriously jeopardise the survival of the more usual berry-eaters. Even mistle thrushes, and they are not small birds in the usual sense, are afraid to go on to a tree while the pigeons are there. They sit on the bare branches of nearby trees, flicking their wings in frustration, and quickly fly off if a pigeon so much as flutters up into the air to get a better foothold on the drooping branches.

Since a score of woodpigeons takes several days to strip a 20 foot holly, we have an indication of the abundance in a berry crop over the countryside as a whole, for the capacity of a pigeon's crop is almost unbelievable. It can hold 1,000 wheat grains, 60 acorns or 20 small potatoes.

SAVED BY THE SCRAPS

The redwing, a regular winter visitor here, spends its summers in northern Europe. It is about the size of a song thrush and much the same colour, perhaps slightly darker. There are, however, two marked differences. It has chestnut-red flanks, very noticeable, as it flies up, and a pronounced buffish-white stripe over the eye.

Usually it feeds in loose flocks in fields, but is perhaps more familiar as a solitary bird raking among the dead leaves under trees and shrubberies looking for worms, slugs, snails and insects. But when snow covers the ground and it cannot get through it has to look elsewhere for its food.

In hard winters we can often see how the woodpigeons, having eaten down the green vegetables, are driven to stripping the hawthorns and hollies of their berries. It could well be that this might jeopardise the survival of smaller birds. There is a sequel, a nice point of ecology.

The pigeons, untidy feeders, leave the ground under each tree littered with fallen berries. Redwings, although not habitual berry-eaters, can be quick to find these, and I have seen a tight flock of 30 or more follow from one hawthorn or holly to another in the wake of the pigeons, day after day, clearing every fallen berry.

It is easy enough to condemn a bird or animal that conflicts with our interests, and the ravages of woodpigeons are not easy to condone. There can, however, be little doubt that the redwings, desirable visitors in every respect, have been saved from starvation by the maligned woodpigeon.

CHAPTER 12

HIBERNATION

WAKENED BY COLD?

The sleep of hibernating animals, like our own nightly sleep, varies from deep to light, so that sometimes the hibernant wakes up and becomes active. Bats sleeping in caves may move from one roost to another, or even fly into the open on a frosty night.

Even the occasional hedgehog can be seen foraging in a hard frost, although the usual comment is that this happens only when the weather is mild. A Dorset reader saw one, active and perky and far from sleepy, one night during a spell of very hard weather, and saw a hedgehog, presumably the same one, the following night. She wonders whether the brightness of the snow gave the illusion of moonlight to tempt the hedgehog out.

There is, perhaps, a more likely explanation. The main difference between the sleep of warm-blooded hibernants and our nightly sleep is that the hibernant relinquishes its temperature control, so that its body temperature falls to a very low level and the animal behaves temporarily as if cold-blooded. Should the temperature of the surrounding air drop below that of its own body, which has fallen from the normal 35.1° C to 6 or 7°, the hedgehog responds by becoming once again warm-blooded and fully active. This is not so remarkable as it sounds, because even when deep in its winter sleep the blood in the hedgehog's heart is still at 35° C and only at the surface of the body is it 6–7° C.

MURDERED IN BED

During the summer dormice enjoy greater security than most

113

small animals. They can only rarely be seen about by day, and normally they sleep out the daylight hours. Their sleeping nests are in the forks of saplings, usually well above ground level, not readily accessible to marauders on the ground and not particularly vulnerable to winged enemies.

At night, when feeding, they clamber through the slender branches, above the reach of ground predators, and protected by twigs and foliage from owls. One of their main dangers is from

Hibernating dormouse

domestic cats, some of which will travel as much as three miles each way in a night to hunt in the woods.

In October the dormouse hibernates and we think of it securely tucked up in its nest, in a hollow stump, in a hole in the ground, or more commonly under the carpet of leaf litter. If this notion were correct there would be the puzzle why this small animal, leading a life of little danger, should be so rare.

The answer seems to be that winter is its danger time. Far from being safely asleep it has come down to ground level, and its fibrous nest, excellent for keeping out cold, is no protection against marauders. Predatory birds like magpies and carrion crows find it. Foxes, badgers, stoats and weasels, at times possibly even rats, may between them kill four out of every five during this resting period.

OLD ERROR ABOUT SQUIRRELS

It is many years now since Barrett-Hamilton, writing on the red squirrel in his standard work on British mammals, referred to 'the still more erroneous belief that the animal remains during the grea-

Grey squirrel

ter part of the winter in a state of almost complete torpidity'. Yet even today one hears it said confidently, often by those who should be better informed, that squirrels go to sleep for the winter. There are several surprising aspects to the persistence of this error.

The first is that the old belief should quite readily have been transferred from the native red squirrel to the imported grey. The second is that those who believe the squirrel hibernates usually believe also that it stores food for use in winter, without noticing the implied contradiction. The third point is that every authoritative writer on the subject denies the hibernation legend. As a test, I recently looked into more than a score of popular books on British wildlife, and in each the writer described the belief as erroneous.

Finally, squirrels are common enough everywhere, either the red or the grey, and it should be a matter of common observation that one or more can be seen running about on even the bleakest of cold days in winter, and their tracks are probably the commonest of all wild animal tracks after a fall of snow. Doubtless they sleep longer hours in winter, and stay more indoors, but don't we all?

THE CUCKOO'S THREE MISSING MONTHS

'I am wondering whether a few cuckoos shirk the sea-crossing and skulk around hidden in the woods in winter.' So asks a correspondent, who says: 'We hear a great deal about the people who claim to hear the first cuckoo. I feel I could claim to have seen him this year, on March 13.'

For 2,000 years there have been claims that some birds hibernate. In this country the old idea was that cuckoos turned into hawks in winter and that swallows spent it in the mud at the bottom of the ponds. Gilbert White tells of a clergyman who reported finding swifts asleep in the rubble of a church tower that was being demolished in winter. All these stories used to be rejected. Then in 1946 a poor-will, the North American nightjar, was actually found hibernating. Since then hibernation—with an almost complete cessation of breathing, lowered temperature and apparently lifeless condition—has been proved in the case of the whip-poor-will, another nightjar, and several species of swifts and nearly a dozen other birds are suspected of hibernation.

The earliest record in this country for the cuckoo is 10 March, the latest 26 Dec. For the swift, the dates are 21 March and 21 Dec. What happens to these birds during the intervening three months?

The hibernating poor-will was first found by the merest accident after 2,000 years of scoffing. A £10 note to whoever finds the first hibernating cuckoo—as well as fame!

HOT AND COLD BATS

By about April bats are coming out of winter quarters and taking up summer roosts. At such times they do odd things. Pipistrelles can be seen hunting at midday in full sunlight or in early evening with a chill wind. It is as if the bats are perplexed, not having settled yet into their normal rhythm.

One reason why their habits are not easy to understand is that bats are both warm-blooded and cold-blooded. In a truly cold-blooded animal the body temperature varies more or less with the temperature of the surrounding air. A warm-blooded animal can exercise control of its body temperature, keeping it more or less constant.

Bats in hibernation relinquish their temperature-control. The temperature of the body falls to that of the air in the hibernaculum, the pulse-rate and breathing drop to the point where they are almost imperceptible. The bat looks all but dead.

Something very similar happens in the daytime during the months of spring and summer when the bats are active.

In most warm-blooded animals asleep the body temperature is only a degree or two lower than normal. The daily slumber of a bat during summer is very near to hibernation. The temperature-control is relinquished, the body grows cold, the pulse and breathing rate drop. Then, towards the evening, as the time for active hunting draws near, the bat slowly comes up out of its deep, cold-blooded slumber to become warm-blooded again.

FROG BENEATH THE ICE

In the pond immediately in front of my house, the frogs come out of hibernation around mid-February and soon start breeding.

This same pond gives the unusual opportunity to see where some at least of the frogs spend the winter.

There is no doubt that many hibernate in the ground, because they can be seen at the appropriate time coming out and making their way overland. It is also suggested that some hibernate in the mud at the bottoms of ponds, but this is less easy to see or to prove.

One December, a few days before Christmas, the frogs in my pond suddenly appeared during a mild spell. There were 11 of them, varying from small to large, and they might have come from the ground surrounding the pond or from the mud in it. Then came the freeze-up, and the frogs could be seen swimming under 3 inches of ice. As the days passed their numbers dwindled until only 5 large females were left. They were obviously in breeding condition and presumably had the urge to stay out. When the ice finally melted they were dead. The cold had been too much for them.

The rest had disappeared and, because the sheet of ice had covered the whole surface of the pond without a crack or an exit of any kind, they must have gone into the mud at the bottom and stayed there.

ANTS CAUGHT NAPPING

A clod of earth turned out by the spade contained a mass of ants several layers deep and 3 to 4in across. So this is how they spend the winter. To one side of the mass was a patch like white-looking eggs; closer inspection showed it to be ant grubs, partly rolled up.

Either because they had been disturbed or because the air was warmer at the surface, a few of the ants soon started to move, slowly and lethargically. Those still in the mass were also stretching their legs a little, in very slow motion. Their bodies were numbed and their senses clogged by winter sleep. Those that had detached themselves from the compact mass began to wander about, slowly and erratically. But instinct will out and very soon each picked up one of the white grubs in its mouth.

A colony of ants is practically all female, the workers being sterile females. Perhaps this is why their first reaction when a nest is disturbed is to look to the safety of the next generation, an instinct so strong that this is almost the first act when rudely aroused.

To the other side of the mass of ants was another white cluster: miniature woodlice of a species that habitually lives in ants' nests. Each was milk white, translucent and about one-sixteenth of an inch long. They also were asleep, not mingling with the ants but there ready to join the nest when full activity is resumed in spring.

A peacock butterfly, frequently found hibernating in the house, perhaps in the folds of the curtains

HIBERNATING FLIES

Houseflies pester us all the summer and then disappear for the winter—but not quite. Occasionally they are found in large numbers hibernating in houses. Other kinds of flies spend the summer out-of-doors but seek sheltered spots to hibernate as adults.

The most habitual offenders are the clusterflies, slightly larger than a housefly and with brassy-coloured hairs on the thorax. They are often seen in large clusters, especially in attics, but except that groups of any flies are unattractive they are harmless, and if their presence is considered undesirable they can be killed by the usual spraying.

That, however, will probably be the end only for one year. Any of the flies that hibernate in buildings tend to occupy the same sites year after year. This is not like swallows coming back each year to the same barn to nest. There can be no 'homing' instinct in the flies because those we see this winter will be dead and replaced by another generation next winter. It can only be supposed that it is something about the conditions offered by a particular room that attracts them.

We tend to associate hibernating flies with old houses, but new houses are not exempt. In a recent letter I was told of a block of flats completed last April which was invaded by scores of bluish-green flies, slightly smaller than the more familiar greenbottles. These settled on the windows behind the curtains. Obviously this was no return to traditional winter quarters.

One of these invading hibernants is a small black and yellow fly. It is an occasional visitor indoors. This should be spared as there is reason to believe that its larvae feed on aphides or greenfly.

SNAKES AND ADDERS

One reads of scores of snakes, hibernating in a tangled mass, being dug out of a bank in winter. An unfulfilled ambition of mine is to see such a thing, but it was partly fulfilled in the first week of one December when my good friend the woodman brought me an adder. He had found it quiescent under a piece of bark lying on the ground beneath dead bracken.

Although obviously torpid there were sufficient signs of movement to check any tendency on our part to take liberties with it. We supposed that because there had been no very hard weather up to then the hibernation was not complete.

Then in January the woodman brought along a grass snake, found under bracken, but although we had by now had a spell of snow this snake also was merely quiescent. Since a grass snake is non-venomous we were more disposed to handle it and look closely at it. It was limp and showed only slight movements of the muscles. When fully active it would be darting its tongue rapidly in and out. Now, as we handled it, it was doing so languidly and sleepily.

The hibernation of a cold-blooded animal is vastly different from that of a warm-blooded animal, such as a hedgehog, which is preceded by many physiological changes. Cold-blooded animals merely become torpid, and the current practice is to speak of their entering not hibernation but a winter rigidity. Judging by our adder and grass snake even this is something of a misnomer.

CHAPTER 13

MIGRATION

EARLY CUCKOOS

By the second week of April cuckoos have been heard everywhere. Although the main stream, in common with the majority of our bird visitors from the south, does not arrive until April, solitary birds may reach us before that. 10 March is the earliest record, and from then on arrivals will increase in numbers, reaching a peak in the first weeks of April, and then tailing off again.

Migration is usually presented to us as a sudden departure of thousands of birds from one place to another, over a distance of a few miles or thousands of miles, according to the species. But it is never as clear-cut as this.

Migration is not merely a matter of taking wing and travelling with timetable regularity along a well-defined route. These things are only the outward and visible signs of an inward upwelling. The more obvious features of the act of migration are baffling to explain, and we are even more in the dark over the physiological changes that prompt it.

When the seasonal rhythm sets in, a bird hitherto content with a limited range will spread its wings and depart for a distant territory, impelled by forces that are the synthesis of changes in almost every organ of its body and guided by senses we have only begun to guess at. In such a concatenation we should expect individual variation, so some cuckoos are early, some late. After all, even on the best-run railways the trains are not always on time.

SWALLOWS READY FOR DEPARTURE

In autumn swallows start to congregate on the telegraph wires.

We see them morning after morning, and again in the evenings, their numbers growing as the time wears on. Then one morning they are there no longer; they have set off on the long trek to South Africa.

It is a familiar story, its details household knowledge. There is, however, one point worthy of comment: the element of rehearsal.

Swallow at nest

We are prone to attribute many things to man's reasoning behaviour which are, in fact, almost universal in the animal kingdom. The trick of rehearsal is one of them. We use it as a deliberate and purposive thing, planned and rational. But there are many day-to-day things which, quite unnoticed by ourselves, we rehearse before they are set in motion. At some time in the development of the human race, the use of rehearsals was lifted from an instinctive to an intuitive level, and thence put on a rational basis. And now we are conceited about it.

123

Man has invented little that did not exist in some natural form long before he inherited the earth. Swallows, and other things, have been using the techniques of rallying points, rehearsals and zero hour for departure for a very, very long time. The baffling thing is that they manage to carry out these manoeuvres with such precision and without, so far as we can see, leaders, words of command or any other of the many advantages we ourselves possess.

MARTINS' INDIAN SUMMER

By October it seems a long time since house-martins were seen gathering on the telegraph wires, and doubtless most of them have reached their winter quarters in South Africa. In the south of England, however, groups of six to a dozen may still be about, flying around during days of bleak weather no less actively than in warm autumn sunshine. Neither weather nor temperature seems to be a decisive influence on migration.

The main arrival of house-martins begins in the south in the first week in April, but individual birds have been seen as early as 8 March. From early April the influx continues until the first week in June. Not long after this the first signs of the return migration are seen. About mid-July the first birds to return move towards the south coast, and in the first week in August emigration begins in earnest. It continues until the third week in October, or even mid-November. Exceptionally, it may last into December, and the latest records of all are for 22 Dec for a house-martin seen in Suffolk and 10 Jan for one seen in Middlesex.

NIGHTINGALES' MEMORY

Not far away a Nightingale Corner is now embodied in an arterial road. In the other direction Nightingale Lane is a concreted road serving a new building estate. Nightingale this and Nightingale that occur as local names in southern England with a frequency greater than that accorded other bird-names. This may be due to the popularity of the bird's song, or to its localised habitat.

In our woods the nightingales yearly return to roughly the same clumps of bushes. There is one in the south-west corner and one to the north-east. Nearby patches of undergrowth, and others lying between the two widely separated sites, seem as suitable for nesting sites.

If we may judge nightingales by what is known of other migratory species, half the old birds that leave us in September will not return, and only half this year's youngsters will survive to come back. It appears also that young or old return to the ancestral site.

It is remarkable enough that swallows should come back each year to the same barn, but a barn does not alter in appearance as vegetation does. To return from winter quarters in tropical Africa to the same bushes suggests a fine memory of landmarks. But then memory itself is a remarkable thing, and of tremendous importance in all animals.

SPIDERS GO FLYING

A reader sent me this spider mystery: 'A few mornings ago we found a long spider's thread which stretched from the top of the garage to an overhanging tree branch, a distance of about 15ft. How was the thread carried across by the spider? There was a low bush on route but the thread did not touch this in any way. The spider could hardly have precipitated itself across the space.'

On a fine sunny morning in autumn, when the air is fairly still, currents of warm air rise from the ground. At such times, spiders climb up the grass stems, turn their bodies into such wind as there is, and let out a thread of silk. This is caught by the moving air and pulled out, without any assistance from the spider. During the last week or two I have watched several of these migratory flights. As the rays of the sun caught the threads I could see them rising obliquely from the grass, some 20 or more feet long. After a while each would float away with its spider at the end. Perhaps one such thread landed between the garage and the bough.

The young of many kinds of spiders scatter in this way during the summer, but these autumn migrations are made by adult spiders of a particular kind. Being of small size they can use this form of aerial transport, although why they do so is not known. Perhaps it is to prevent overcrowding.

Besides the dearth of butterflies in the unusual summer of 1959, many readers noticed an exceptional number of moths, including some rarities. There was a real abundance, especially of hawk-moths. Readers reported seeing the rare Oleander Hawk-moth, the Death's-head Hawk-moth, and the Convolvulus Hawk-moth, and the scarce Olive-tree Pearl was seen in the south and west. Above all, the year was remarkable for the numbers of the Humming-bird Hawk-moth. On 8 Sept, at the end of one of the hottest days, a Newquay reader saw some 5,000 moths on a bank of valerian, mainly the Silver-Y moth but also 'an incredible number of Humming-bird Hawk-moths'.

This sudden appearance and disappearance of insects has been noted for centuries, but only within the past half century has the idea been accepted that it is due to migration. We now know, beyond question, that some moths, notably hawk-moths, and many butterflies not only reach us from long distances but travel

Humming-bird hawk-moths at a flower

right across Britain to the Orkneys and Shetlands, and actually make the return flight in autumn.

I have several times seen swarms of the Large White butterfly heading across the Channel for the English coast and have followed them up through the southern counties. They have the purposeful appearance of migratory flights of birds. The numbers involved can be gauged by an accident in 1911, when a swarm settled on a sundew-covered island in Sutton Broad in Norfolk and 6 million were trapped. In years of heavy migrations our resident populations of this butterfly are swollen to plague proportions.

Many of the migrant moths and butterflies, however, are unable to stand the winter here, and but for their migratory habit we should not have the pleasure of seeing them in our gardens. The Humming-bird Hawk-moth is found in southern Europe and North Africa, and ranges also across Asia, even to Japan, but many thousands migrate north each year. There seems to be a return flight in the autumn, although solitary individuals have been recorded here even in winter.

Why these insects migrate is a question to which there is as yet no answer. At first it was supposed they represented an overflow from the breeding areas, but the deliberate nature of the migration makes this unlikely. Nor can it be an escape from enemies, since the columns of migrating butterflies are harried by insect-eating birds all along the route.

How do insects find their way and keep direction? They will remain on course whether they have a following wind, a cross wind or a head-on wind. Bad weather may halt the flight temporarily, but it makes no difference to the route followed, nor does the amount of sunshine. Prolonged study of this problem has failed to give an answer.

DANGEROUS CROSSING

There is a road near here which toads cross each spring on their way to the breeding-pond. They come in from due east, drop down the bank on one side of the road and scramble up the bank on the other side, to reach their destination. Their migration takes place by night, but their passage is marked by day by the bodies flattened under car-wheels.

In late October the autumn migration starts. By night, in the beam of a torch, toads can be seen dropping down the bank on one side, crossing the road and climbing the opposite bank, in the reverse direction. Their route is across arable land, which makes difficult attempts to follow it farther than the road, but presumably they scatter and dig into the ground somewhere for hibernation.

From other sources we know there is a spring migration to a breeding-pond, that the toads spend the summer in the neighbourhood of that pond and then return in autumn along the same route to their winter quarters.

Although that is the overall pattern the details are hard to fill in for any given area, largely because the journeys take place at night. We know the toads converge on the pond along several routes and that they navigate by the stars. The breeding-ponds are easy to find. Their hibernating sites are usually discovered only by accident. The pitiful corpses on the road tell us, if nothing else, how precise are the migration routes. They are always in roughly the same spot, spring and autumn.